AF770448

Growth Hormone and the Heart

ENDOCRINE UPDATES
Shlomo Melmed, M.D., Series Editor

---

1. E.R. Levin and J.L. Nadler (eds.): Endocrinology of Cardiovascular Function. 1998. ISBN: 0-7923-8217-X
2. J.A. Fagin (ed.): Thyroid Cancer. 1998. ISBN: 0-7923-8326-5
3. J.S. Adams and B.P. Lukert (eds.): Osteoporosis: Genetics, Prevention and Treatment. 1998. ISBN: 0-7923-8366-4.
4. B.-Å. Bengtsson (ed.): Growth Hormone. 1999. ISBN: 0-7923-8478-4
5. C. Wang (ed.): Male Reproductive Function. 1999. ISBN 0-7923-8520-9
6. B. Rapoport and S.M. McLachlan (eds.): Graves' Disease: Pathogenesis and Treatment. 2000. ISBN: 0-7923-7790-7.
7. W. W. de Herder (ed.): Functional and Morphological Imaging of the Endocrine System. 2000. ISBN 0-7923-7923-9
8. H.G. Burger (ed.): Sex Hormone Replacement Therapy. 2001. ISBN 0-7923-7965-9
9. A. Giustina (ed.): Growth Hormone and The Heart. 2001. ISBN 0-7923-7212-3

Growth Hormone And The Heart

Editor

Andrea Giustina, M.D.

Assistant Editor

Filippo Manelli, M.D.

Endocrine Section
Department of Internal Medicine
University of Brescia, Italy

KLUWER ACADEMIC PUBLISHERS
BOSTON/DORDRECHT/LONDON

Distributors for North, Central and South America:
Kluwer Academic Publishers
101 Philip Drive
Assinippi Park
Norwell, Massachusetts 02061 USA
Telephone (781) 871-6600
Fax (781) 871-6528
E-Mail <kluwer@wkap.com>

Distributors for all other countries:
Kluwer Academic Publishers Group
Distribution Centre
Post Office Box 322
3300 AH Dordrecht, THE NETHERLANDS
Telephone 31 78 6392 392
Fax 31 78 6546 474
E-Mail <orderdept@wkap.nl>

Electronic Services <http://www.wkap.nl>

Library of Congress Cataloging-in-Publication Data

Growth hormone and the heart / editor, Andrea Giustina; assistant editor, Filippo Manelli.
p. ; cm.—(Endocrine updates; 9)
Includes bibliographical references and index.
ISBN 0-7923-7212-03 (hardback: alk. paper)
1. Somatotropin—Physiological effect. 2. Somatotropin—Pathophysiology. 3. Heart—Diseases—Endocrine aspects. I. Giustina, Andrea. II. Manelli, Filippo. III. Series.
[DNLM: 1.Heart Diseases—physiopathology. 2. Somatotropin—physiology. 3. Heart—physiology. 4. Somatotropin—therapeutic use. WG 210 G884 2001]
QP572.S6 G7584 2001
616.1'207—dc21

00-046037

Copyright © 2001 by Kluwer Academic Publishers.

All rights reserved. No part of this publication may be reproduced, stored in a retrieval system or transmitted in any form or by any means, mechanical, photo-copying, recording, or otherwise, without the prior written permission of the publisher, Kluwer Academic Publishers, 101 Philip Drive, Assinippi Park, Norwell, Massachusetts 02061

Printed on acid-free paper.

Printed in the United States of America

The Publisher offers discounts on this book for course use and bulk purchases. For further information, send email to barbara.murphy@wkap.com.

Table of Contents

List of Contributors

Maria R. Ambrosio:
Department of Biomedical Sciences and Advanced Therapies
(Section of Endocrinology), University of Ferrara,-Italy.

Stefan D. Anker
Franz-Volhard-Klinik- (charité, Campus Berlin-Buch) at
Max Delbruck-Centrum for Molecular Medicine, Berlin, Germany.

Bengt-Åke Bengtsson:
Research Center for Endrocinology and Metabolism
Sahlgrenska University Hospital, Goteborg, Sweden.

Micaela Bernareggi:
School of Medicine, University of Milano-Bicocca, Milano, Italy

Ferruccio Berti:
School of Medicine, University of Milano-Bicocca, Milano, Italy.

Elena Bresciani:
Department of Pharmacology, Universitv of Milan, Italy.

Francesco Cavagnini:
Chair of Endocrinology, University of Milan, IRCCS Ospedale San Luca,
Istituto Auxologico Italiano, Via Spagnoletto n. 3, I-20149 Milan, Italy.

Philippe Chanson:
Service d'Endocrinologie et des Maladies de la Reproduction, University
Hospital Bicetre, Assistance Publique, Hopitaux de Paris and Faculté de
Médicine Paris-Sud, Université Paris X, Kremlin Bicetre, France.

Jens S. Christiansen:
Department of Endocrinology and Diabetes, Aarhus University Hospital,
Kommunehospitalet, DK-8000 Aarhus C, Denmark.

Mariantonietta Cicoira:
Cardiac Medicine, NHLI, Imperial College School of Medicine, London, UK.

Antonio Cittadini:
Department of Internal Medicine and Cardiovascular Sciences, University
Federico II, Naples, Italy.

Annamaria Colao:
Department of Molecular and Clinical Endocrinology and Oncology,
Frederico II University of Naples, Italy.

Lisa K. Conley:
College of Agriculture, Forestry and Life Sciences, Clemson University, Clemson, South Carolina, USA.

Vito De Gennaro Colonna:
Departinent of Pharmacology, University of Milan, Italy.

Romano Deghenghi:
Europeptides Argenteuil, France

Ettore C. Degli Uberti:
Department of Biomedical Sciences and Advanced Therapies (Sect-ion of Endocrinology), University of Ferrara, Italy.

Marc Y. Donath:
Division of Endocrinology and Diabetes, Department of Medicine., Universitv Hospital, Zurich, CH.

Pamela S. Douglas:
Beth Israel Deaconess Medical Center, Harvard Medical School, Boston, MA, USA.

Serafino Fazio:
Department of Internal Medicine and Cardiovascular Sciences, University Federico II, Naples, Italy.

Paola Franceschetti:
Departinent of Biomedical Sciences and Advanced Therapies, (Section of Endocrinology), University of Ferrara, Italy.

Andrea Giustina:
Endocrine Section, Department of Internal Medicine, University of Brescia, Italy.

Ken K.Y. Ho:
Pituitary Research Unit, Garvan Institute of Medical Research, St. Vincent's Hospital, Sydnev, Australia.

Jorgen Isgaard:
Research Center for Endocrinology and Metabolism, Sahlgrenska University Hospital, Goteborg, Sweden.

Grace Y.W. Kam:
Pituitary Research Unit, Garv Institute of Medical Research, St. Vincent's Hospital, Sydney, Australia.

Vittorio Locatelli:
School of Medicine, University of Milano-Bicocca, Milano, Italy.

Gaetano Lombardi:
Department of Molecular and Clinical Endocrinology and Oncology, "Federico II" University of Naples, Italy.

Roberto Lorusso:
II Cardiac Surgery, Spedali Civili di Brescia, I-25125 Brescia, Italy.

Marco Losa:
Department of Neurosurgery, IRCCS San Raffaele, Via Olgettina 60, I-20132 Milano, Italy.

Filippo Manelli:
Endocrine Section, Department of Internal Medicine, University of Brescia, I-25125 Brescia, Italy.

Eugenio E. Muller:
Department of Pharmacology, University of Milan, Italy.

Claus-Dieter Pflaum:
Medizinische Klinik, Ludwigs-Maximilian Universität, Munich, Germany.

Angela I. Pincelli:
Chair, Endocrinology, University of Milan, IRCCS Ospedale San Luca, Istituto Auxologico Italiano, Milan, Italy.

Giuseppe Romanelli:
Geriatric Section, Department of Internal Medicine, University of Brescia, Italy.

Giuseppe Rossoni:
Department of Pharmacology, University of Milan, Italy.

Luigi Saccà:
Department of Internal Medicine and Cardiovascular Sciences, University Federico II, Naples, Italy.

Massimo Scacchi:
Chair of Endocrinology, University of Milan, IRCCS Ospedale San Luca, Istituto Auxologico Italiano, Milan, Italy.

Christian J. Strasburger:
Medizinische Klinik, Ludwigs-Maximilian Universität, Munich, Germany.

Leif Thuesen:
Department of Endocrinology and Diabetes, Aarhus University Hospital, Kommunehospitalet, Aarhus, Denmark.

Antonio Torsello:
School of Medicine, University of Milan-Bicocca, Milan, Italy.

Mary Lee Vance:
Division of Endocrinology and Metabolism, University of Virginia Health Sciences Center, Box 601, Charlottesville, VA 22908, USA.

Johannes D. Veldhuis:
University of Virginia School of Medicine, General Clinical Research Center, Charlottesville, VA 22908, USA.

Maurizio Volterrani:
Fondazione "S. Maugeri", Department of Cardiology, Gussago Medical Center, I-25125 Brescia, Italy.

Klaus Von Werder:
Endocrinology—Dept. of Medicine, Schlosspark Klinik, Humboldt University, Berlin, Germany.

William B. Wehrenberg:
College of Agriculture, Forestry and Life Sciences, Clemson University, Clemson, SC USA.

Arthur Weltman:
University of Virginia School of Medicine, General Clinical Research Center, Charlottesville, VA 22908, USA.

Troels Wolthers:
Pituitary Research Unit, Garvan Institute of Medical Research, St. Vincent's Hospital, Sydney, Australia.

Jürgen Zapf:
Division of Endocrinology and Diabetes, Department of Medicine, University Hospital Zürich, CH-8091 Zürich, Switzerland.

Preface – *Riccardo Vigneri*

The endocrine system is the major regulator of cardiac function: cathecolamines and thyroid hormones, for instance, play an important and well known role for the regulation of the cardiovascular function. In the last 10 years growth hormone and its peripheral mediator IGF-1 have also gained more and more attention for their action on the cardiovascular system. Today, large evidences demonstrate that the GH/IGF-1 axis has great relevance for the regulation of cardiac growth, structure and function.

The mechanisms explaining GH effects on the cardiovascular system are multiple, complex and not yet fully understood. There is certainly a *direct action* of GH and IGF-1 on the heart, through the stimulation of specific receptors for both GH and IGF-1 in the cardiac myocytes, where they promote cell growth; these hormones also affect the cardiovascular stromal tissue biology. The GH/IGF-1 system has also an *indirect effects* on the heart, mainly a consequence of the peripheral vasodilatation. Particularly, IGF-1 stimulates endothelial nitric oxide synthesis and, by this mechanism, causes endothelium-dependent vasodilatation which change myocardial hemodynamic work and cardiac performance.

This book provides an useful update of current understanding of the mechanisms of GH action on the heart. Moreover, it discusses the clinical use of GH, its stimulators and its inhibitors on the heart structure and function, from two different points of view. First, different qualified groups discuss the consequences of chronic GH deficiency or excess on heart function. In patients with either GH deficiency or excess an impaired cardiovascular function occurs in the long-term and may reduce life expectancy. The benefits of treatments that normalize hormone concentrations are also discussed. Second, the use of GH as an agent for treatment of chronic heart failure is discussed by researchers that have discovered and proposed this particular use of GH. This is a field of great interest, with a lot of unsolved questions concerning the dose and duration of the therapeutic use of GH in patients with severe cardiomyopathy, with the possibility of emerging adverse effects when treatment is prolonged.

As already mentioned, the complexity of the relationship between the GH/IGF-1 system and cardiac performance is still incompletely understood. This book will provide the reader with updated information regarding the biology and pathology of different effects of GH/IGF-1 on the heart. In addition, I believe that it will stimulate more interest and more research on areas of possible GH use like miocardial ischemia and heart energetic, both of great interest for possible clinical applications.

Introduction – *Andrea Giustina and Filippo Manelli*

Several hormonal systems regulate heart function in physiological and pathological states. These notions have been organized in a new research and clinical discipline which can be defined "cardiovascular endocrinology". However, that the growth hormone (GH) – IGF-1 system may be heavily involved in the regulation of cardiac structure and function, as suspected on the basis of several clinical observations from many years, has been demonstrated only very recently by solid experimental and clinical evidences. This book aims at structuring in a single publication knowledge that has been so far accumulated in the area of GH and the heart from basic to clinical studies, by research groups working on this topic throughout the world. Lessons from different experimental models and from several human diseases (acromegaly, adult GH deficiency, heart failure) suggest to both endocrinologists and cardiologists that not only GH may have a role in the physiology and pathophysiology of heart function but that GH itself may find a place in the treatment of primary heart diseases (such as dilated cardiomyopathy) or of cardiac complications of hypopituitarism.

The influences that GH has on heart function are also important for understanding the potential role of GH in improving exercise performance. Therefore, we thought that it of interest for the many people who are interested in the study of exercise (cardiologists, endocrinologists, physiologists and others) to provide an update of studies of the effects of exercise and nutritional status on GH secretion as well as of GH effects on the skeletal muscle and fluid/electrolyte balance before addressing in its complexity the effects of GH on exercise performance. In fact, the current knowledge seems to support the view that endogenous GH may influence exercise performance not only via its positive effects on the cardiac system but also acting at the level of skeletal muscle, nutritional status and physical training.

Finally, the last section of the book is a window on the role of other components of the somatotropic axis (IGF-1, GH secretagogues) in the regulation of heart structure and function. The role of these molecules is important also for their clinical and therapeutical potential as alternatives to GH in the treatment of heart dysfunction.

In conclusion, we hope that this book will not only be a useful update of the research that has been very actively produced in the field of cardiovascular endocrinology but also the primary step in the recognition of the wide physiological and clinical significance of GH and the heart interactions.

Growth Hormone and the Heart

1
THE CELLULAR AND MOLECULAR BASIS FOR GROWTH HORMONE ACTION ON THE HEART

Antonio Cittadini and Pamela S. Douglas*
*University Federico II, Naples, Italy, and *Beth Israel Deaconess Medical Center, Harvard Medical School, Boston MA - USA*

INTRODUCTION

Despite a series of elegant studies performed by Beznak in the fifties clearly demonstrating Growth Hormone (GH)'s role in maintaining a normal hypertrophic response following experimental aortic constriction (1-3), the cardiovascular role of the GH/IGF-1 axis was largely overlooked. In fact, until few years ago, GH was classically regarded as the principal mediator of bone linear growth, in addition to its well known actions on glucose homeostasis and skeletal muscle mass, and the heart was not generally considered as a target tissue (4). It is now well established that not only does GH play an essential role in maintaining a normal cardiac muscle mass and function, but also that its growth-promoting and positive inotropic actions may be beneficial in the setting of heart failure and ischemic syndromes.

The heart is a complex organ whose regulation has sequentially appeared during evolution, first by controlling the synthesis of gene products, then by controlling the internal cell environment, and finally by controlling the organ as a whole, as a specialized tissue (5). From this vantage point, the interaction between GH and the heart can be examined by shifting emphasis among three levels: at the molecular level (i.e. synthesis of myocardial proteins), at the level of cell biochemistry and biophysics (i.e. calcium fluxes and cardiac contractility), and at the whole organ level (i.e. loading conditions and pump function). This chapter will dwell upon the molecular and cellular interactions between GH and the heart.

GH and IGF-1 secretion and binding proteins

The precise mechanism(s) governing GH and IGF-1 secretion are still a matter of debate. The somatomedin hypothesis states that GH acts via a principal mediator insulin-like gowth factor-1 (IGF-1), whose hepatic secretion is under GH control, and then reaches the target tissues through the blood stream, as a classical endocrine factor (6). This view was supported by the demonstration that IGF-1 can mimic the growth-promoting actions of GH in hypophysectomized rats (7). However, this hypothesis was challenged in the mid-eighties by the documentation of multiple sites of synthesis of IGF-1, in particular in the heart, whose extractable tissue concentration in normal rats was lower than liver, lung, and kidney, but higher than muscle, brain, testes etc (8). It was therefore concluded that also paracrine/autocrine mechanism(s) play a prominent role in GH actions. This is consistent with the general view that peptide growth factors, at variance with classic hormones, are produced by many different tissue and exert their effects on neighboring tissues (paracrine effects) or even on their own cells of origin (autocrine effects). To reconcile such different roles of GH and somatomedins, Green et al., studying a preadipocyte cell line, proposed the so called dual effector model, suggesting that locally produced IGF-1, under GH regulation, contributes to the stimulatory effects of GH (9). It is also well accepted that GH may have direct actions on some target tissues, independent of IGF-1.

Circulating IGF concentrations in humans are about 1000 times higher than those of insulin, but unlike insulin, most of the biological activity of IGF in serum is restrained by IGF binding proteins (10). Of IGF-1 binding proteins, IGFBP-3 is the most abundant in serum. It has been recently suggested that IGFBP-3 may play a role in modulating cardiac function, since, when administered alone to ferret papillary muscle, it decreased isometric developed tension in a dose-dependent fashion with a maximal reduction of 36% (11). Concomitant administration of IGFBP-3 and IGF-1 at equimolar concentrations blocked IGF-1 positive inotropic and growth-promoting effects. A similar finding was previously reported by Ito and colleagues, who reported that IGFBP-3 induced a dose-dependent inhibition of protein synthesis stimulated by IGF-1's in cultured rat cardiomyocytes (12). Taken together, it could be speculated that endogenous IGF-1 production does exist in the heart, and that it may help in maintaining a normal level of contractility. Moreover, it is possible that an interaction between IGFBP-3 and IGF-1 may also play a role in the regulation of cardiac hypertrophy.

MOLECULAR INTERACTIONS

Numerous peptide growth factors have been so far identified in myocardial tissue (13). However, it is unclear whether most of them exert significant biological actions on the heart under normal circumstances. In this regard, there are some fundamental prerequisites that a certain peptide needs to possess to support its role as a candidate molecule for myocardial growth and hypertrophy, such as: 1) presence of receptors within the myocardial tissue; 2) presence of the growth factor at the protein or the mRNA level within the myocardial tissue; 3) change of the growth factor myocardial content following the variation of the hemodynamic load; 4) direct effects of the growth factor on the isolated heart preparation and 5) significant cardiovascular actions following its exogenous acute or chronic administration.

Several papers published mainly in the last decade have unequivocally proven that IGF-1, the principal mediator of GH's effect, possesses such prerequisites (14). In fact, the presence of GH and IGF-1 receptors in the heart and specifically in myocyte membranes has been identified by several independent groups (15-17). Moreover, it has been shown in animal models that cardiac volume or pressure overload increase IGF-1 myocardial content (18-20). Such increases occur particularly in the left ventricle, and not in the right ventricle, probably in relation to the different wall stresses (18). Very recently, Neri Serneri et al. have documented in human myocardium enhanced synthesis of IGF-1 following both pressure (aortic stenosis) and volume (aortic regurgitation) overload (21). Such IGF-1 mRNA overexpression was localized in the cardiomyocyte compartment and, interestingly, had a biphasic response related to the degree of elevation of wall stress. Direct effects of IGF-1, but not GH, on the isolated heart and/or cardiomyocytes have been demonstrated by a number of authors and include: increased protein synthesis (12), augmented contractility and changes in calcium handling (11,22-24). Moreover, in isolated cardiomyocytes, IGF-1 induces an early and sustained expression of the muscle specific genes for troponin I and myosin light chain-2 that parallels the development of cell hypertrophy (12). Finally, exogenous administration of GH and IGF-1 in normal rats and in rats with experimental heart failure encompasses a broad variety of cardiovascular effects in addition to the expected somatic growth, ranging from activation of myocardial growth, peripheral vasodilation, and changes in wall stress and cardiac function (25-29). In humans, reactivation of cardiac growth and peripheral vasodilation represent the most consistent consequences of GH administration, while modification of cardiac function is less clear (30-32).

Intracellular pathways

Generally speaking, it is well accepted that the growth factor signaling cascade is extremely complex, with more than 500 genes regulated by numerous diverging, cross-talking, and redundant pathways. The intracellular pathway(s) activated by GH and IGF-1 binding to their cognate receptors are just beginning to be investigated (Figure 1).

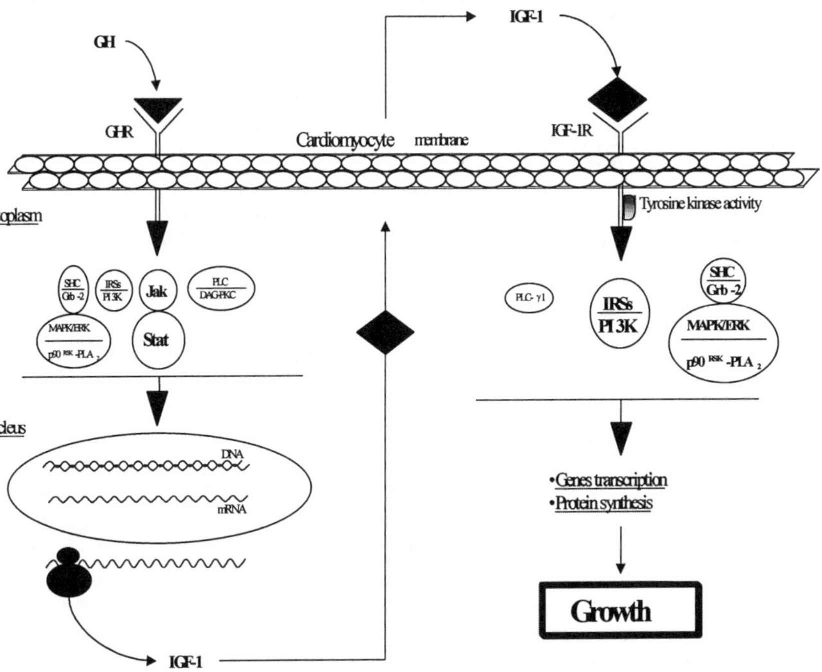

Figure 1. Possible scheme of the intracellular pathways activated by GH and IGF-1 binding to their receptors.

GH stimulates the tyrosine phosphorylation of various cellular polypeptides, including the GH receptor itself, in an early part of the intracellular response (33). It also possesses tyrosine kinase activity. These phosphorylations are catalyzed by JAK2, a GH receptor-associated kinase member of the Janus family of tyrosine kinases (34). Moreover, GH also activates MAP kinases, STAT transcription factors, and has been recently shown to increase the tyrosine phosphorylation of IRS-1, IRS-2, and Shc in several target tissues, including the heart (35).

Also the signal transduction system of IGF-1 in cardiac cells has recently begun to be elucidated. Several major substrate proteins have been recently demonstrated to be activated by IGF-1 binding to its receptor including IRS-1, the Shc proteins, and Crk, a cellular homologue of v-crk (36). IRS-1, in turn, binds the p85 β-subunit domain of PI3-kinase (37). In this concern, it has been recently demonstrated that the PI3-kinase pathway is activated by IGF-1 in rat cardiomyocytes (38) and is involved in IGF-1-induced prevention of apoptosis in PC12 cells (39). In addition to the activation of IRS-1, Shc, and Crk, other second messengers may be generated by IGF-1 receptor activation, including the phospholipase C pathway. In fact, IGF-1 stimulation leads to a rapid accumulation of inositol phosphates and corresponding increases in cytoplasmic Ca^{++} in thyroid cells (38,40). Activation of phospholipase C and rapid elevation of intracellular inositol phosphate levels have been reported by Foncea et al. and Guse et al. (38,40). Other intracellular pathways activated by IGF-1 in cardiac cells include c-raf, ERK1 and ERK2 (with maximal activation after 5-minute IGF-1 exposure of 8.3 and 12-fold, respectively,), p90 S6 kinase, with a concomitant stimulation of [^{3}H]phenylalanine (1.6 fold), without any effect on [^{3}H]thymidine incorporation (38).

In addition to myocardial growth, cardiac function can also be potentially affected by the activation of these intracellular systems. In fact, the phosphoinositidase second-messenger cascade involves bifurcated end products, among which protein kinase C activation has been recently shown to modulate postive inotropic responses in rat ventricular myocytes (41). Intracellular alkalosis, that follows protein kinase C activation, also sensitizes the myofilaments to Ca^{++} (42).

CELLULAR INTERACTION

Contractility

There is now robust evidence that the activation of the GH/IGF-1 axis may change intrinsic contractility independent of its growth promoting action. In fact, using isolated heart preparations, specifically cardiomyocytes, Vetter et al. (22) and subsequently Freestone and collegues (23) showed that IGF-1 infusions evoked a dose-dependent increase of cell shortening in isolated cardiomyocytes. A more recent study by Cittadini et al. confirmed such findings also in the isolated perfused heart and in the papillary muscle (11). The reported increase of contractility in the various studies was about 20-25% from baseline values. The concentration response curve of IGF-1 demonstrated an initial effect at at about 10^{-8}, and a maximal response at 10^{-7}, the usual ligand-receptor concentration range of most organic compounds. The response was fast, starting only few seconds after IGF-1 application to the perfusion media, and did not appear to be species-specific, being evident in rat, ferret, and mouse heart.

The IGF-1-induced increase of intrinsic contractility has also been investigated in disease conditions such as heart failure, diabetes mellitus, hypertension, or aging (43,44). In diabetes mellitus, hypertension, and aging, there is a marked reduction of IGF-1 positive effects, a sort of "IGF-1 resistance" whose pathophysiological role has not been clarified, but appear linked to altered NO production (43,44). Conversely, in myocytes isolated from dogs with pacing-induced heart failure, Kinugawa et al. recently demonstrated the preservation of the ability of IGF-1 to increase contractility, which was similar to that observed in myocytes from normal animals (24).

Interestingly, GH does not appear to exert any direct contractile effect on the isolated heart or papillary muscles or cultured cardiomyocytes (11,12). It is therefore plausible that the physiological role of GH receptor in the heart does not include acute modulation of myocardial contractility.

Calcium handling

All studies looking at IGF-1-induced acute changes in contractility consistently demonstrate a positive inotropic response following IGF-1 application, independent of the model system and of the animal species employed. However, there are conflicting results as to the subcellular mechanism (s) underlying such effects, particularly concerning calcium fluxes during excitation-contraction coupling.

Changes in calcium handling could be considered as the final common pathway of all interventions that modify myocardial contractility. Viewed from this perspective, the contractile state of cardiac muscle can be altered by the following mechanisms:"upstream" mechanisms that alter the amplitude or time-course of the Ca^{++} transient or alter the affinity of troponin C for Ca^{++} or "downstream" mechanisms that alter the response of the myofilaments to a given level of occupancy of the Ca^{++} binding sites on troponin C (42). Freestone and coworkers (23) and Kinugawa et al. (24), found an increase of intracellular Ca^{++} in isolated cardiomyocytes subjected to acute IGF-1-application, paralleling the significant increase of contractility, and whose magnitude might account entirely for the observed changes of contractility (i.e. "upstream" mechanisms). Conversely, Cittadini et al. found no significant changes of intracellular Ca^{++} activator available to the contractile machinery when IGF-1 was added to the perfusate, but an increase of Ca^{++} responsiveness of the myofilaments (11). This finding was obtained under steady-state tetanized conditions, in isovolumic buffer perfused aequorin-loaded isolated heart preparations.

In chronic settings, the activation of myocardial growth is also associated with changes of myofilament composition (25,45), and the direct action of IGF-1 on calcium fluxes cannot be dissociated from those secondary to myocardial growth. Also such studies have yielded conflicting results. Mayoux and coworkers, in rats bearing a GH-secreting tumor, found a 39% increase of the maximal Ca^{++} activated force per cross-sectional area in muscles exposed to GH excess compared with control muscles (45). This finding was associated with a slight increase of Ca^{++} sensitivity of the contractile proteins, with unchanged cross-cycling rate, and with a significant decrease of α-myosin heavy chain. Also Strömer et al. found an augmented Ca^{++} responsiveness of the myofilaments, due to a significant increase of maximal activated Ca^{++} force, but a slight decrease in the EC_{50} of the Ca^{++} concentration curve (46). In this study, also intracellular Ca^{++} concentrations were measured by means of the aequorin method, and no differences were found between GH and placebo treated rats. Congruent results were recently obtained in ventricular myocytes from mice constitutively overexpressing IGF-1 by Redaelli et al, who reported a positive influence of IGF-1 on cardiac performance, due to enhancement of shortening velocity and cellular compliance, and a slight decrease of Ca^{++} sensitivity of the myofilaments (47). A recent study performed in rats with experimental myocardial infarction has shown that chronic treatment with GH for 2 weeks increases the cardiac contractile reserve assessed by the peak Ca^{++} - cell shortening relationship in fluo 3 loaded isolated LV myocytes (48). Specifically, at higher Ca^{++} concentrations in the perfusate, GH was able to increase toward control values the intracellular Ca^{++} transients, that were significantly depressed in rats with myocardial infarction. This finding was associated with a significant increasse of SERCA2 mRNA and protein levels in the left ventricle of GH-treated rats. In this regard, another mechanism by which GH hypersecretion may enhance cardiac contractility is by prolonging the action potential, and therefore increasing calcium influx through the L-type channels (i.e. "upstream" mechanisms). To support this speculation, Xu et al. demonstrated decreased transient outward K^+ current in acromegalic rats (49).

It appears difficult to reconcile these findings. In general, no doubt exists that the activation of the GH/IGF-1 axis is associated with positive inotropy (i.e. increase of myocardial contractility), but discrepancies exist as to the precise changes in Ca^{++} handling. Such inconsistencies are likely multifactorial, and might be explained by differences of Ca^{++} indicator, animal species, and experimental conditions. For instance, it is known that Ca^{++} sensitivity of the myofilaments varies in isotonic vs. isometric conditions, in steady-state vs. twitching muscles etc.

GH/IGF-1 axis and cardiac molecular phenotypes

It is well known that the development of LV hypertrophy and failure in humans and animal models is associated with major changes of structural protein, excitation contraction coupling, and energy metabolism (50). In addition, there is the reexpression of a fetal gene program including, among others, the upregulation of β-myosin heavy chain, atrial natiuretic factor, lactate dehydrogenase M subunits, B subunit of creatin kinase, and by downregulation of β_1-adrenergic receptors, and myoglobin (50). Sarcolemmal proteins such as SERCA2, phospholamban, and ryanodine receptor have been shown to be downregulated both at the protein and at the mRNA level in a broad variety of hypertrophy/failure models. On the other hand, the sarcolemmal Na- Ca^{++} exchanger, that is supposed to compensate for the downregulation of SERCA2 with the attendant intracellular Ca^{++} overload, has been reported to be increased.

GH appears to play a prominent role in the regulation of these molecular adaptations. Myosin composition is changed by GH excess, with a significant shift toward the slow V3 isoform, along with accelerated synthesis of troponin I and myosin light chain 2. GH directly stimulates the expression of SERCA2, that is the most important regulator of Ca^{++} uptake by the sarcoplasmic reticulum. In rats with postinfarction heart failure, SERCA2 is increased by GH treatment (48). Similar findings were observed in the genetic model of the cardiomyopathic hamster (51). GH treatment attenuated the decrease of ryanodine receptor density and affinity, and this was associated with the prevention of the development of the heart failure phenotype. Finally, unpublished results from our laboratory aimed at comparing GH-induced cardiac growth and pathologic hypertrophy due to aortic banding in the rat, have shown that GH treatment does not alter the molecular pattern of Ca^{++} regulating genes, while a similar extent of myocardial growth due to aortic constriction induces the typical phenotype changes described (Isgaard J, Cittadini A, Strömer H, personal observations).

Peripheral vasodilation

One of the most consistent and relevant biological effects of GH is peripheral vasodilation (Figure 2). It is thought to be mediated by IGF-1, which possesses high affinity binding sites on endothelial cells. IGF-1 has a direct nitric oxide (NO) releasing effect, recently shown in cultured human endothelial cells (52). Subsequently, this finding has been confirmed in a forearm model in healthy humans, in which IGF-1 infusion induced significant vasodilation which was reversed by the addition of L-NMMA (53). NO in turn is a known paracrine mediator of vasodilation, inhibitor of platelet aggregation, leukocyte adhesion, and of smooth muscle cell growth (54,55). The subcellular mechanisms of IGF-1 action of smooth muscle cells involve the stimulation of the Na^+-K^+ pump, followed by reduction of intracellular Na^+ and the activation of the Na^+-Ca^{++} exchanger in the reverse mode, which compensates for Na^+ exit from the intracellular milieu. This leads to a decrease of Ca^{++} with attendant muscle relaxation and vasodilation. On the other hand, IGF-1 also increases NO release in smooth muscle cells, with consequent augmentation of its second messenger cyclic GMP, and reduction of intracellular Ca^{++} (54,55).

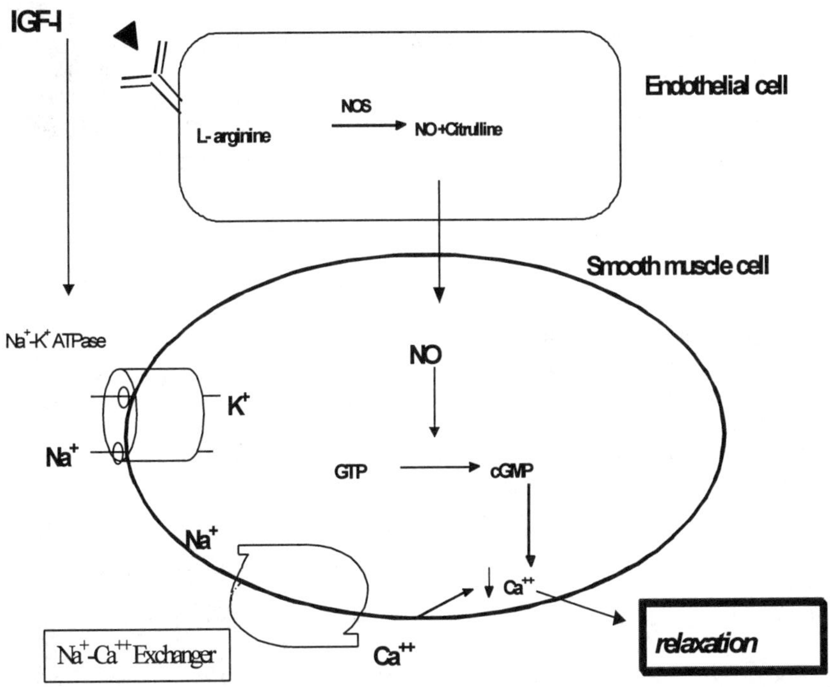

Figure 2. Schematic diagram of IGF-1 effects on endothelial and smooth muscle resulting in vasodilatation.

Energetics

A pioneering study by Timsit et al. first suggested that GH-induced cardiac growth displays unique characteristics from an energetic standpoint. In fact, despite a significant phenoconversion toward the V3 myosin isoform, there was an increase of the maximum isometric force of papillary muscle normalized per cross-sectional area (25). In other words, GH allowed the muscle to improve its contractile performance and economy simultaneously, a unique pattern of myocardial adaptation. This was in all likelihood due to an increased number of active enzymatic sites and to the concomitant myosin phenoconversion.

A similar way to increase performance without increasing energetic expenditure was also observed during IGF-1 application to isolated heart preparations, since 20-25% increase of contractility was associated with no change of intracellular ATP content and high energy phosphates (11). Such observations, in association with the decrease of systolic wall stress, may provide the cellular basis for the dramatic decrease of oxygen consumption observed in seven patients with idiopathic dilated cardiomyopathy treated for 3 month with recombinant GH (56). Whether GH might modify the expression of candidate proteins for myocardial energetics in normal or diseased conditions remains to be elucidated.

CONCLUSIONS

One hundred years after the first description of cardiac involvement in acromegaly by Huchard (57), there is now solid clinical evidence for an prominent role of the GH/IGF-1 axis in cardiovascular physiology. In this chapter, we have provided the molecular and cellular basis for such role. Future research is needed to better clarify the interactions between GH and the heart and to use this knowledge in devising threapeutic approaches.

REFERENCES

1. Beznak M. The effect of the pituitary and growth hormone on the blood pressure and on the ability of the heart to hypertrophy. J Physiol 1952;116:74-83.
2. Beznak M. The effect of adrenocortical hormones alone and in combination with growth hormone on cardiac hypertrophy and blood pressure of hypophysectomized rats. J Physiol 1954;124:75-83.
3. Beznak M. The restoration of cardiac hypertrophy and blood pressure in hypophysectomized rats by large doses of lyophilized anterior pituitary and growth hormone. J Physiol 1954;124:64-74.
4. Thorner MO, Vance ML, Horvath E, Kovacs K. The anterior pituitary. In: Wilson JD, Foster DW (eds) Williams Textbook of Endocrinology (8th edition). Philadelphia: WB Saunders Company. 1992:221-310.
5. Katz AM. Regulation of myocardial contractility. In: Katz AM (ed) Physiology of the heart. New York: Raven Press; 1992:319-350.
6. Salmon WD Jr, Daughaday WH. A hormonally controlled serum factor which stimulates sulfate incorporation by cartilagene in vitro. J Lab Clin Med 1957;49:825-836.
7. Guler H-P, Zapf J, Scheiwiller E, Froesch ER. Recombinant human insulin-like growth factor I stimulates growth and has distinct effects on organ size in hypophysectomized rats. Proc Nat Acad Sci 1988;85:4889-4893.
8. D'Ercole AJ, Stiles AD, Underwood LE. Tissue concentrations of somatomedin C: further evidence for multiple sites of synthesis and paracrine or autocrine mechanisms of action. Proc Natl Acad Sci USA 1984;81:935-939.
9. Green H, Morikawa M, Nixon T. A dual effector theory of growth-hormone action. Differentiation 1985;29:195-198.
10. Sara VR, Hall K. Insulin-like growth factors and their binding proteins. Physiol Rev 1990;70:591-614.
11. Cittadini A, Ishiguro Y, Strömer H, et al. Insulin-like growth factor-1 but not growth hormone augments mammalian myocardial contractility by sensitizing the myofilaments to Ca^{2+} through a wortmannin-sensitive pathway: studies in rat and ferret isolated muscles. Circ Res 1998;83:50-59.

12. Ito H, Hiroe M, Hirata Y, et al. Insulin-like growth factor-I induces hypertrophy with enhanced expression of muscle specific genes in cultured rat cardiomyocytes. Circulation 1993;87:1715-1721.
13. Hefti MA, Harder BA, Eppenberger HM, Schaub MC. Signaling pathways in cardiac myocyte hypertrophy. J Mol Cell Cardiol 1997;29:2873-2892.
14. Saccà L, Cittadini A, Fazio S. Growth hormone and the heart. Endocr Rev 1994;15:555-573.
15. Mathews LS, Enberg B, Norstedt G. Regulation of rat growth hormone receptor gene expression. J Biol Chem 1989;17:9905-9910.
16. Engelmann GL, Bohem KD, Haskell JF Khairallah, Ilan J. Insulin-like growth factors and neonatal cardiomyocyte development: ventricular gene expression and membrane receptor variations in normotansive and hypertensive rats. Mol Cell Endocrinol 1989;63:1-14.
17. Sklar MM, Kiess W, Thomas CL, Nissley SP. Developmental expression of the tissue insulin-like growth factor II/mannose 6-phosphate receptor in the rat. J Biol Chem 1989;264:16733-16738.
18. Isgaard J, Wählander H, Adams MA, Friberg P. Increased expression of growth receptor mRNA and insulin-like growth factor-I mRNA in volume overloaded hearts. Hypertension 1994;23:884-888.
19. Wählander H, Isgaard J, Jennische E, Friberg P. Left ventricular insulin-like growth factor I increases in early renal hypertension. Hypertension 1992;19:25-32.
20. Hanson MC, Kenneth AF, Alexander RW, DeLafontaine P. Induction of cardiac insulin-like growth factor I gene expression in pressure overload hypertrophy. Am J Med Sci 1993;306:69-74.
21. Neri Serneri GG, Modesti PA, Boddi M, et al. Cardiac growth factors in human hypetrophy:relations with myocardial contractility and wall stress. Circ Res 1999;85:57-67.
22. Vetter U, Kupferschmid C, Lang D, Pentz S. Insulin-like growth factors and insulin increase the contractility of neonatal rat cardiocytes in vitro. Bas Res Cardiol 1988;83:647-654.
23. Freestone NS, Ribaric S, Mason WT. Mol Cell Biochem. 1996;163/164: 223-229.
24. Kinugawa S, Tsutsui H, Ide T, et al. Positive inotropic effect of IGF-1 on normal and failing cardiac myocytes. Cardiovascular Research 1999;43:157-164.
25. Timsit J, Riou B, Bertherat J, et al. Effects of chronic growth hormone hypersecretion on intrinsic contractility, energetics, isomyosin pattern, and myosin adenosine triphosphatase activity of rat left ventricle. J Clin Invest 1990;86: 507-515.
26. Duerr RL, Huang S, Miraliakbar HR, Clark R, Chien KR, Ross JJr. Insulin-like growth factor-1 enhances ventricular hypertrophy and function during the onset of experimental cardiac failure. J Clin Invest 1995;95:619-627.
27. Cittadini A, Strömer H, Katz SE, et al. Differential cardiac effects of growth hormone and insulin-like growth-factor-1 in the rat: a combined in vivo and in vitro evaluation. Circulation 1996;93:800-809.
28. Cittadini A, Grossman JD, Napoli R, et al. Growth hormone attenuates early left ventricular remodeling and improves cardiac function in rats with large myocardial infarction. J Am Coll Cardiol 1997;29:1109-1116.
29. Duerr RL, McKirnan D, Gim RD, Clark R, Chien KR, Ross JJr. Cardiovascular effects of insulin-like growth factor-I and growth hormone in chronic left ventricular failure in the rat. Circulation 1996;93:2188-2196.

30. Thuesen L, Christiansen JS, Sorensen KE, Jorgensen JOL, Orskov H, Henningsen P. Increased myocardial contractility following growth hormone administration in normal man. Danish Med Bull 1988;35:193-196.
31. Caidahl K, Eden S, Bengtsson B-A. Cardiovascular and renal effects of growth hormone. Clin Endocrinol 1994;40:393-400.
32. Cittadini A, Ehrnborg C, Longobardi S, et al. Growth hormone activates cardiac growth in normal subjects: a multicenter, double-blind, placebo-controlled study. Endocrine Society's 81st Annual Meeting. June 12-15, 1999; San Diego, California. OR4-2.
33. Campbell GS, Christian LJ, Carter-Su. Evidence for involvement of the growth hormone receptor-associated tyrosine kinase in actions of growth hormone. J Biol Chem 1993;268:74427-7434.
34. Argetsinger LS, Campbell GS, Yang X, et al. Identification of JAK2 as a growth hormone receptor-associated tyrosine kinase. Cell 1993;74:237-244.
35. Thirone ACP, Carvalho CRO, Saad MJA. Growth hormone stimulates the tyrosine kinase activity of JAK2 and induces tyrosine phosphorylation of insulin receptor substrates and Shc in rat tissues. Endocrinology 1999;140:55-62.
36. Rubin R, Baserga R. Insulin-like growth factor-I receptor: its role in cell proliferation, apoptosis, and tumorigenicity. Lab Invest 1995;73:311-33.
37. Myers MG Jr, Backer JM, Sun XJ, et al. IRS-1 activates phosphatidylinositol 3'-kinase by associating with src homology 2 domains of p85. Proc Natl Acad Sci USA 1992;89:10350-10354.
38. Foncea R, Andersson M, Ketterman A, et al. Insulin-like growth factor-1 rapidly activates multiple signal trasduction pathways in cultured rat cardiac myocytes. J Biol Chem 1997;272: 19115-19124.
39. Parrizas M, Saltiel AR, LeRoith D. Insulin-like growth factor 1 inhibits apoptosis using the phosphatidylinositol 3'-kinase and mitogen-activated protein kinase pathways. J Biol Chem 1997;272:154-161.
40. Guse AH, Kiess W, Funk B, Kessler U, Berg I, Gercken G. Idenification and characterization of insulin-like growth factor receptors on adult rat cardiac myocytes: linkage to inositol 1,4,5-triphosphate formation. Endocrinology 1992;130:145-151.
41. Pi Y, Sreekumar R, Huang X, Walker JW. Positive inotropy mediated by diacylglycerol in rat ventricular myocytes. Circ Res 81:92-100.
42. Blinks JR, Endoh M. Modification of myofibrillar responsiveness to Ca^{2+} as an inotropic mechanism. Circulation 1986;73: III85-98.
43. Ren J, Walsh MF, Hamaty M, Sowers JR, Brown RA. Altered inotropic response to IGF-I in diabetic rat heart: influence of intracellular Ca^{2+} and NO. Am J Physiol 1998;275:H823-H830.
44. Ren J, Sowers JR, Natavio M, Brown RA. Influence of age on inotropic response to insulin and insulin-like growth factor I in spontaneously hypertensive rats: role of nitric oxide. Proc Soc Exp Biol Med 1999;221:46-52.
45. Mayoux E, Ventura-Clapier R, Timsit J, Behar-Cohen F, Hoffman C, Mercadier JJ. Mechanical properties of rat cardiac skinned fiber are altered by chronic growth hormone hypersecretion. Circ Res 1993;72:57-64.
46. Strömer H, Cittadini A, Douglas PS, Morgan JP. Exogenously administered growth hormone and insulin-like growth factor I alter intracellular Ca^{++} handling and enhance cardiac performance. Circ Res 1996;79:227-236.

47. Redaelli G, Malhotra M, Li B, et al. Effects of constitutive overexpression of insulin-like growth factor-1 on the mechanical characteristics and molecular properties of ventricular myocytes. Circ Res 1998;82:594-603.
48. Tajima M, Weinberg EO, BartunekJ, et al. Treatment with growth hormone enhances contractile reserve and intracellular calcium transients in myocytes from rats with postinfarction heart failure. Circulation 1999;99:127-134.
49. Xu X, Best PM. Decreased transient outward K^+ current in ventricular myocytes from acromegalic rats. Am J Physiol 1991;260:H935-H942.
50. Swynghedauw B. Molecular mechanisms of myocardial remodeling. Physiol Rev 1999;79:215-262.
51. Takeshi U, Ohkusa T, Yano M, Matsuzaki M. Growth hormone preserves cardiac sarcoplasmic reticulum Ca^{2+} release channels (ryanodine receptors) and enhances cardiac function in cardiomyopathic hamsters. Cardiovasc Res 1998;40:64-73.
52. Tsukahara H, Gordienko DV, Tönshoff B, Gelato MC, Gloigorsky MS. Direct demonstration of insulin-like growth factor-I-induced nitric oxide production by endothelial cells. Kidney Int 1994;45:598-604.
53. Fryburg DA. N-monomethyl-L-arginine inhibits the blood flow but not the insulin-like response of forearm muscle to IGF-I. Possible role of nitric oxide in muscle protein synthesis. J Clin Invest 1996;97:1319-1328.
54. Böger RH. Role of nitric oxide in the haemodynamic effects of growth hormone. Growth Horm IGF Res 1998;8:163-165.
55. Delafontaine P. Growth factors and vascular smooth muscle cell growth responses. Eur Heart J 1998;19:G18-G22.
56. Fazio S, Sabatini D, Capaldo B, et al. A preliminary study of growth hormone in the treatment of dilated cardiomyopathy. N Eng J Med 1994;334:809-814.
57. Huchard H. Anatomie pathologique, lesions et troubles cardiovasculaires de l'acromegalie. J Praticiens 1895;9:249-251.

2
PHYSIOLOGICAL EFFECTS OF GROWTH HORMONE ON THE HEART

Giovanni Lombardi and Annamaria Colao
Department of Molecular and Clinical Endocrinology and Oncology
"Federico II" University, Naples, Italy

INTRODUCTION

In recent years, GH was recognized not only to regulate the somatic growth but also to profoundly affect several physiological processes such as the maintenance of muscle mass and strength, body composition and fuel metabolism. GH, both directly and *via* its peripheral effector IGF-I, is involved in the development and maintenance of heart structure and function (1,2). In fact, some clinical and experimental studies reported that either GH excess or deficiency almost always include an impaired function of the left ventricle, which may stepwise lead to manifest heart disease, thus reducing life expectancy (1,2). This finding is related both to metabolic changes such as hyperlipidemia, increase of body fat and premature atherosclerosis and to a primitive impairment of the heart. As GH was demonstrated to induce an increase in the left ventricle mass, the potential application of GH treatment in idiopathic dilated cardiomyopathy and in congestive heart failure has also been recently investigated with controversial results (3-5). The influence of the GH/IGF-I axis on the heart is displayed at different levels (Fig.1). In this chapter recent knowledge on GH/IGF-I effect on the heart are reviewed.

CONCEPTS OF NORMAL CARDIAC PHYSIOLOGY

The ability of the heart to function as a pump is measured as cardiac output or the volume of blood ejected by the left ventricle per minute. Cardiac output is determined by the product of the heart (beats/minute) and stroke volume. Stroke volume, in turn, is determined by pre-load (the amount of blood presented to the heart to be pumped), after-load (the forces opposing ejection of the blood from the ventricle), myocardial contractility (the intrinsic ability of the heart to develop force) and diastolic function (the ease with which blood can enter the ventricle). While each of the determinants of stroke volume exerts an independent effect on cardiac function, there is an interdependence among these factors as well. Changing one often affects one or more of the others. Heart function is also coupled to the systemic and pulmonary vascular beds as well as to the central nervous system. Diseases and drugs typically exert their effect on more than one of the determinants described above (for a review see ref. 6).

Cardiomyocytes are considered to be terminally differentiated cells thus cardiac growth in adult life, determines hypertrophy mainly and a hypertrophic cardiac response generally occurs as a physiological adaptation to increased hemodynamic load, except in acromegaly (see below).

GH/IGF-I AXIS AND CARDIAC STRUCTURE AND FUNCTION

GH and IGF-I regulate cardiac development and myocardial growth. The GH receptor gene is expressed in the myocardium (7). Rats with surgically removed pituitary glands respond to GH administration either with an increase in cardiac IGF-I content (8), or an increase of the expression of IGF-I mRNA in a dose-dependent manner (9). Cardiac myocytes express IGF-I receptors (10) and IGF-I causes hypertrophy of cultured rat cardiomyocytes, increasing mRNA levels of sarcomeric proteins including myosin light chain-2 and troponin (11). An increase of IGF-I mRNA was found in rat myocardium after pressure overload secondary to either banding of ascending aorta, or aortocaval shunt, myocardial infarction to experimental renal hypertension (12-15). Increased IGF-I immunoreactivity was found in the inner layers of the left ventricle (15) where both tension and wall stress are high with a gradual decline towards the epicardial surface (16). No elevation of IGF-I was observed in the right ventricle (17). The onset of IGF-I expression preceded or paralleled the development of left ventricular hypertrophy (15,18). These results indicate that hypertension and increased left ventricular pressure load constitutes a mechanical stimulus for a local growth response in the ventricle and locally synthesized IGF-I may play a role (17). This hypothesis was also supported by the evidence that increased wall stress may trigger a local induction of IGF-I expression in experimental models of RV challenge, through a surgically induced fistula between abdominal aorta and caval vein (19) or induction of pulmonary hypertension (20). Moreover, the exogenous administration of GH and IGF-I in normal adult rats induced a cardiac hypertrophic response without development of significant fibrosis (21). GH and IGF-I play also a direct effect on

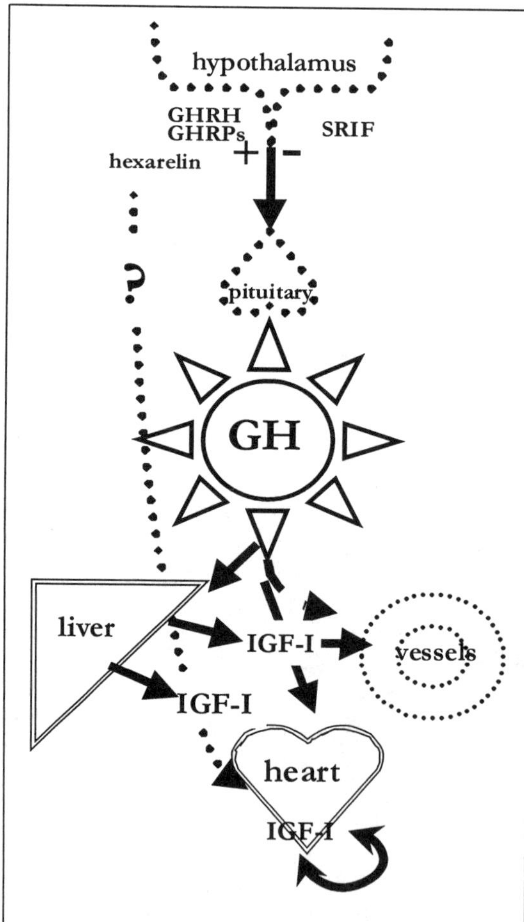

Figure 1. Possible sites of action of the GH/IGF-1 axis on the cardiovascular system.

myocardial contractility. Increase in contractility was shown in preparations of cardiac tissue from animal models of chronic GH excess (22), likely due to increased calcium responsiveness of myofilaments (23). In addition, more recently, acute intravenous infusion or subcutaneous administration of recombinant IGF-I to healthy subjects has been reported to improve cardiac performance assessed by impedance cardiography or echocardiography, respectively (24,25). The intravenous infusion of 60 μg/Kg of IGF-I over 3 h induced systemic side effects in addition to metabolic changes (24) while the same dose given s.c. did not induce side effects or modify glucose levels while inhibiting insulin levels as well as the GH response to physical exercise (25). Moreover, the s.c. injection of 20 μg/Kg of IGF-I was demonstrated to enhance cardiac inotropism in healthy subjects (26). This low "physiological" IGF-I dose did not significantly inhibit spontaneous GH levels while reduced insulin and glucose levels (26). The increase in left ventricular ejection fraction induced by IGF-I could be due either to a direct cardiac effect or to secondary to vasodilatation. In fact, IGF-I displays vasodilator effects in forearm muscles and kidneys (27-29). Another explanation may involve cathecolamine release, although the results reported by Bisi et al. (26), did not show changes in epinephrine nor norephinefrine secretion after rIGF-I administration.

CARDIAC EFFECTS OF GH DEFICIENCY AND EXCESS

Some of the former data available in the Literature concerning the effect of GH on the heart have been drawn from the description of cardiac abnormalities in GH deficiency (GHD) and excess. The evidence that normalization of GH and/or IGF-I after replacement or suppression by surgery or pharmacotherapy, respectively, reverse some cardiac abnormalities, provided further support for a direct role of GH and IGF-I on the heart. Patients with GHD were also reported to have impaired systolic and diastolic function although with a reduced left ventricular mass and (30-33). As in acromegaly, at peak exercise the indexes of left ventricular systolic function remained markedly depressed (30-33). The impairment of the left ventricular performance at rest and on effort was recently demonstrated in a large series of patients with biochemical diagnosis of GHD (peak GH <9 μg/l after arginine plus GHRH test) but without evidence of other complications able to affect heart function, by using equilibrium radionuclide angiography (34). Either at rest or at peak exercise left ventricular ejection fraction was significantly lower in patients than in controls, regardless of age (Fig.2). A normal response of left ventricular ejection fraction to exercise was found only in 28.9% of patients (34). Peak GH, plasma IGF-I levels and disease duration were not correlated to cardiac parameters of left ventricular function neither at rest nor at peak exercise (34). Therefore, impairment of cardiac performance occurs in the large majority of adult patients with GHD but does not seem to be related to disease severity and/or duration. Cardiac performance, but not cardiac mass, was impaired also in elderly GHD patients as in young and adult GHD patients (35). Cardiac impairment often reverse when GH replacement treatment is started. In particular, left ventricular ejection fraction and diastolic filling improve and resting heart rate, wall thickness, exercise tolerance stroke volume increase (30,32,36,37). However, cardiovascular disease in GHD patients is not only due to a direct impairment in cardiac function but also to the associated abnormalities of lipid and carbohydrate metabolism, which contribute to the increased risk for vascular disease.

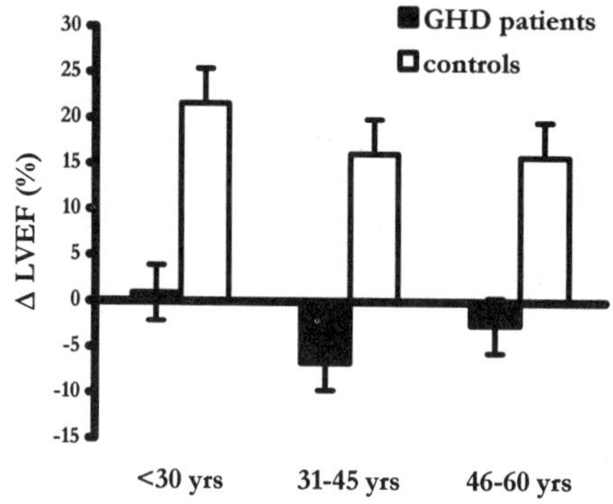

Figure 2. Changes in the left ventricular ejection fraction on effort (Δ LVEF) in very young, young and adult GHD patients.

In symptom-free hypopituitary adult patients, a significant increase in intima-media thickness, a significant reduction in the arterial compliance and distensibility, an increase in the stiffness index and a higher number of atheromatous plaques have been reported as compared to healthy controls (38,39). It has been recently reported that GH treatment in GHD adult patients reverses early morphological and functional atherosclerotic changes in major arteries (40,41). This beneficial effect allows to hypothesized that, if maintained, GH replacement to GHD patients may reduce vascular morbidity and mortality, likely acting *via* IGF-I which is known to have important effects on endothelial cell function (42). In fact, IGF-I has direct releasing effect of nitric oxide, a paracrine mediator of vasodilatation, inhibition of platelet aggregation, and leukocyte adhesion and inhibition of smooth muscle cell growth (43), in cultured human endothelial cells (44). Decreased biological activity of nitric oxide has been shown to be associated with atherosclerosis and hypertension (45,46). In GHD patients a decrease in nitric oxide production responsive to GH treatment has been reported (47). The increased nitric oxide formation after GH replacement may be one of the factors determining improvement of cardiovascular performance in GHD patients (47).

When GH and IGF-I levels increase to supraphysiological levels such as in acromegaly, the involvement of the heart is very common as firstly described by Huchard in 1895 (48). Acromegalic patients show several abnormalities in cardiac morphology and function and peculiarly they have cardiomegaly which is disproportionate as compared to the increase in size of other internal body organs (1,2). Myocardial hypertrophy with interstitial fibrosis, limphomononuclear infiltration and areas of monocyte necrosis results in increased mass and concentric hypertrophy of both ventricles (49-52). Mitral and aortic valve disease, as well as cavity dilatation, may be observed in the late stage. The involvement of the heart in GH excess has a very early onset as signs of cardiac hypertrophy can be revealed also in young acromegalic patients with a presumably short duration of the disease (53,54). Although the presence of other complications such as arrhythmias, hypertension and diabetes mellitus clearly worsen cardiac function in acromegaly (55), a significant impairment of the left ventricular performance was demonstrated also in patients with active but uncomplicated acromegaly (56). In fact, although the left ventricular ejection fraction at rest was normal (>50%) in all but 2 patients and in all controls, the left ventricular ejection fraction at peak exercise was significantly decreased in elderly compared to young patients and to age-matched controls and a normal response of the left ventricular ejection fraction to exercise was found in 30% of patients vs. 87.5% of controls (56). Consequently, exercise-induced changes in left ventricular ejection fraction were significantly decreased both in young and elderly patients as compared to controls (Fig.3).

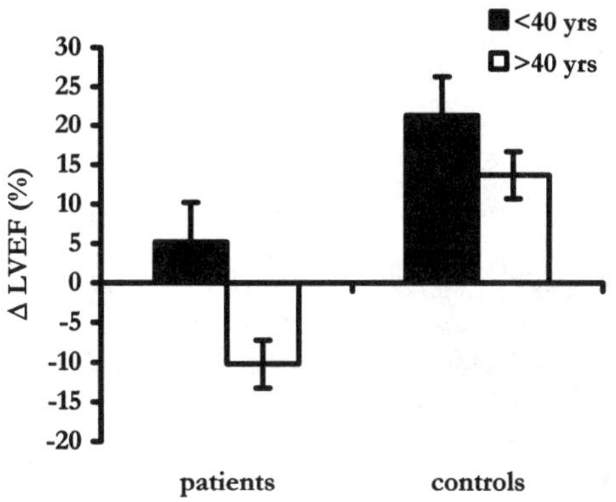

Figure 3: Changes in the left ventricular ejection fraction on effort (Δ LVEF) in young and adult acromegalic patients (modified from ref. 32).

A significant decrease of left ventricular filling was also observed in the patients who presented higher systolic and diastolic blood pressure at peak exercise (56). Age and estimated disease duration were also both significantly correlated with peak rate of left ventricular filling, indicating a direct role of chronic GH excess in inducing cardiac impairment (56,57). At further support to this hypothesis, suppression of GH and IGF-I levels was demonstrated to induce a significant decrease of the left ventricular mass (58-60). Improvement in some indexes of left ventricular diastolic function was observed after lanreotide treatment (54). Furthermore, hormone suppression to safe levels, i.e. GH levels <2.5 μg/l, as fasting sample, or <1 μg/l after glucose load, together with normalization of plasma IGF-I levels, for 1 year was followed by a significant improvement of left ventricular ejection fraction either at rest and at peak exercise without significant changes of diastolic filling (61). Interestingly, the persistence for 1 year of elevated hormone levels caused a significant increase of systolic blood pressure and further impaired cardiac performance (61). These data suggest that a prolonged suppression of circulating GH and IGF-I levels could normalize cardiac performance and likely reverse the poor prognosis for cardiovascular disease in acromegaly (62).

GH SECRETAGOGUES

In recent years several reports underlined the role played by the GH secretagogues (GHS) in the pathophysiology of the GH/IGF-I axis, and more generally, in pituitary secretion (63). GHS are artificial compounds developed to release GH *in vitro*, acting though receptors different from those of GHRH (64). GHS mimic an unknown endogenous factor that activates the GHS receptor at the pituitary and hypothalamus (63). The GHS receptor was recently cloned (65), so demonstrating that GHS is a new physiological system regulating GH secretion together with GHRH and somatostatin (63). These compounds, administered alone or in combination with GHRH, are the most potent and reproducible GH releasers and are considered useful tools for the diagnosis of GHD (63). Beyond the effects on GH secretion, the action of GHS on specific heart diseases is most promising (66). In particular, Hexarelin, is an hexapeptide GHS, which possess specific receptors on cardiomyocytes at partial variance with other GHS (67). Beneficial effects of hexarelin on cardiac performance were reported in rats after experimental myocardial infarction (68) and in perfused hearts from GHD rats (69). A single i.v. injection of hexarelin was recently demonstrated to increase the left ventricular ejection fraction, without any change in mean blood pressure or heart rate, in healthy men (70). Conversely, GH did not modify the left ventricular ejection fraction (70). Whether the cardiac actions of hexarelin are direct through specific receptors or indirect through GH/IGF-I axis stimulation is presently not known.

REFERENCES

1. Saccà L, Cittadini A, Fazio S. Growth hormone and the heart. Endocr Rev 1994;15:555-73.
2. Lombardi G, Colao A, Cuocolo A, et al. Cardiological aspects of growth hormone and insulin-like growth factor-1. J Pediatr Endocrinol Metab 1997;10:553-560.
3. Fazio S, Sabatini D, Capaldo B, et al. A preliminary study of growth hormone in the treatment of heart failure in patients with dilated cardiomyopathy. N Engl J Med 1996;334:809-814.
4. Volterrani M, Desenzani P, Lorusso R, D'Aloia A, Manelli F, Giustina A. Hemodynamic effects of intravenous growth hormone in congestive heart failure. Lancet 1997;349:1067.
5. Bengtsson BA, Bergh C, Caidahl K, Lomsky M, Hjalmarson A, Isgaard J. An overview of cardiac function and implication for growth hormone: cardiomyopathy. J Endocrinol Invest 1998;21:80-83.
6. Dreyer WJ, Garson A Jr, Mayer DC, Neish SR. Normal cardiac physiology: a background for understanding the cardiac effects of growth hormone. J Endocrinol Invest 1998;21:46-54.
7. Mathews LS, Enmberg B, Norstedt G. Regulation of rat growth hormone receptor gene expression. J Biol Chem 1989;17:9905-10.
8. Flyvbjerg A, Jorgensen KD, Marshall SM, Orskov H. Inhibitory effect of octreotide on growth hormone-induced IGF-I generation and organ growth in hypophysectomized rats. Am J Physiol 1991;260:E568-74.
9. Isgaard J, Nilsson A, Vickman K, Isaksson OGP. Growth hormone regulates the level of insulin-like growth factor-I mRNA in rats skeletal muscle. J Endocrinol 1989;120:107-12.
10. Guse AH, Kiess W, Funk B, Kessler U, Berg I, Gercken G. Identification and characterization of insulin-like growth factor receptors on adult rat myocytes: linkage to inositol-1,4,5-trisphosphate formation. Endocrinology 1992;130:145-51.
11. Ito H, Hiroe M, Hirata Y, et al. Insulin-like growth factor-I induces hypertrophy with enhanced expression of muscle specific genes in cultured rat cardiomyocytes. Circulation 1993;87:1715-21.
12. Hanson MC, Kenneth AF, Alexander RV, Delafontaine P. Induction of cardiac insulin-like growth factor-I gene expression in pressure overload hypertrophy. Am J Med Sci 1993;306:69-74.
13. Isgaard J, Wahlander H, Adams MA, Friberg P. Increased expression of growth hormone receptor mRNA and insulin-like growth factor-I mRNA in volume overloaded hearts. Hypertension 1994;23:884-888.
14. Reiss K, Meggs LG, Li P, Olivetti G, Capasso JM, Anversa P. Up-regulation of IGF-I, IGF-I receptor and late growth-related genes in ventricular myocytes after infarction in rats. J Cell Physiol 1994;158:160-168.
15. Wahlander H, Isgaard J, Jennische E, Friberg P. Left ventricular insulin-like growth factor-I increases in early renal hypertension. Hypertension 1992;19:25-32.

16. Mirsky J. Elastic properties of the myocardium: a quantitative approach with physiological and clinical applications. In: Handbook of Physiology. Am Physiol Soc I, 1979;14, 2:497-531.
17. Isgaard J, Tivesten A, Friberg P, Bengtsson B-A. The role of the GH/IGF-I axis for cardiac function and structure. Horm Metab Res 1999;31:50-54.
18. Guron G, Friberg P, Wickman A, Brantsing C, Gabrielsson B, Isgaard J. Cardiac insulin-like growth factor I and growth hormone receptor expression in renal hypertension. Hypertension 1996;27:636-342.
19. Isgaard J, Wahlander H, Adams MA; Friberg P. Increased expression of growth hormone receptor mRNA and insulin-like growth factor I mRNA in volume overloaded hearts. Hypertension 1994;23:884-888.
20. Russel-Jones DL, Leach RM, Ward JPT, Thomas CR. Insulin-like growth factor-I gene expression is increased in the right ventricle hypertrophy induced by chronic hypoxia in the rat. J Endocrinol 1993;10:99-102.
21. Cittadini A, Stromer H, Katz SE, et al. Differential cardiac effects of growth hormone and insulin-like growth factor-I in the rat. A combined in vivo and in vitro evaluation. Circulation 1996;93:800-9.
22. Timsit J, Riou B, Bertherat J, et al. Effects of chronic growth hormone hypersecretion on intrinsic contractility, energetics, isomyosin pattern, and myosin adenosine triphosphatase activity of rat left ventricle. J Clin Invest 1990;86:507-515.
23. Strömer H, Cittadini A, Douglas PS, Morgan JP. Exogenously administered growth hormone and insulin-like growth factor-1 alter intracellular Ca^{2+} handling and enhance cardiac performance. In vitro evaluation in the isolated isovolumic buffer-perfused rat heart. Circ Res 1996;79:227-235.
24. Russel-Jones DL, Bates AI, Unmpleby AM, et al. A comparison of the effect of IGF-I and insulin on glucose metabolism, fat metabolism and the cardiovascular system in normal human volunteers. Eur J Endocrinol 1995;25:403-411.
25. Donath MY, Jenni R, Brunner HP, et al. Cardiovascular and metabolic effects of insulin-like growth factor-I at rest and during exercise in humans. J Clin Endocrinol Metab 1996;81:4089-4094.
26. Bisi G, Podio V, Valetto MR, et al. Radionuclide angiocardiographic evaluation of the cardiovascular effects of recombinant human IGF-I in normal adults. Eur J Endocrinol 1999;140:322-327.
27. Guler HP, Schmid C, Zapf J, Froesch ER. Effect of recombinant insulin-like growth factor I on insulin secretion and renal function in normal human subjects. Proc Natl Acad Sci 1989;86:2868-2872.
28. Hirschberg R, Kopple JD. Evidence that insulin-like growth factor-I increases renal plasma flow and glomerular filtrationr ate in fasted rats. J Clin Invest 1989;83:326-330.
29. Copland AC, Nair KS. Recombinant human insulin-like growth factor-I increases forearm blood flow. J Clin Endocrinol Metab 1994;79:230-232.
30. Cittadini A, Cuocolo A, Merola B, et al. Impaired cardiac performance in GH deficient adults and its improvement after GH replacement. Am J Physiol 1994;267:E219-225.
31. De Boer H, Blok GJ, Van Der Veen E. Clinical aspects of growth hormone deficiency in adults. Endocr Rev 1995;16:63-86.
32. Cuocolo A, Nicolai E, Colao A, et al. Improved left ventricular function after growth hormone in patients with hypopituitarism. Assessment with radionuclide angiography. Eur J Nucl Med 1996;23:390-394.

33. Merola B, Cittadini A, Colao A, et al. Cardiac structural and functional abnormalities in adult patients with growth hormone deficiency. J Clin Endocrinol Metab. 1993;77:1658-61.
34. Colao A, Di Somma C, Cuocolo A, et al. Does the age of onset of growth hormone deficiency affect cardiac performance? A radionuclide angiography study. Clin Endocrinol 2000; In Press.
35. Colao A, Cuocolo A, Di Somma C, et al. Impaired cardiac performance in elderly patients with growth hormone deficiency. J Clin Endocrinol Metab 1999;84:3950-3955.
36. Amato G, Carella C, Fazio S, et al. Body composition, bone metabolism, heart structure and function in growth hormone deficient adults before and after growth hormone replacement therapy at low doses. J Clin Endocrinol Metab 1993;77:1671-76.
37. Carrol PV, Christ ER. Growth hormone deficiency in adulthood and the effects of growth hormone replacement : a review. J Clin Endocrinol Metab 1998;83 :382-395.
38. Markussis V, Beshyah SA, Fisher C, Sharp P, Nicolaides AN, Johnston DG. Detection of premature atherosclerosis in symptom-free hypopituitary adults by high resolution ultrasonography. Lancet 1992;340:1188-1192.
39. Markussis V, Beshyah SA, Fisher C, Parker KH, Nicolaides AN, Johnston DG. Abnormal carotid arterial wall dynamics in symptom-free hypopituitary adults. Eur J Endocrinol 1997;136:157-164.
40. Pfeifer M, Verhovec R, Zizek B, Prezelj J, Poredos P, Clayton RN. Growth hormone (GH) treatment reverses early atherosclerotic changes in GH-deficient adults. J Clin Endocrinol Metab 1999;84:453-457.
41. Borson-Chazot F, Serusclat A, Kalfallah Y, et al. Decrease in carotid intima-media thickness after one year growth hormone (GH) treatment in adults with GH deficiency. J Clin Endocrinol Metab 1999;84:1329-1333.
42. Beshyah SA, Johnston DG. Cardiovascular disease and risk factors in adults with hypopituitarism. Clin Endocrinol 1999;50:1-15.
43. Radomski MW, Salas E. Nitric oxide-biological mediator, modulator and factor of injury: its role in the pathogenesis of atherosclerosis. Atherosclerosis 1995;118: S69-S80.
44. Tsukahara H, Gordienko DV, Tonhoff B, Gelato MG, Goligorsky MS. Direct demonstration of insulin-like gorwth factor-I-insuced nitric oxide production by endothelial cells. Kidney Int 1994;45:598-604.
45. Boger R, Bode-Boger SM, Mugge A, et al. Supplementation of hypercholesterolemic rabbits with L-arginine reduces the vascular release of superoxide anions and restores NO production. Atherosclerosis 1995;117:273-284.
46. Kung CF, Lucher TF. Different mechanisms of endothelial dysfunction with aging and hypertension in rat aorta. Hypertension 1995;25:194-200.
47. Boger RH, Skamira C, Bode-Boger S, Brabant G, Von Zur Muhlen A, Frolich JC. Nitric oxide may mediate the hemodynamic effects of recombinant growth hormone in patients with acquired growth hormone deficiency. J Clin Invest 1996;98:2706-2713.
48. Huchard H. Anatomie pathologique, lesions et trouble cardiovasculaires de l'acromegalie. J Practiciens 1895;9:249-250.
49. Fazio S, Cittadini A, Sabatini D, et al. Evidence for biventricular involvement in acromegaly: a Doppler echocardiographic study. Eur Heart J 1993;14:26-33.
50. Hradec J, Marek J, Kral J, Janota T, Polonieski J, Malik M. Long-term echocardiographic follow-up of acromegalic heart disease. Am J Cardiol 1993;72:205-210.

51. Morvan D, Komajda M, Grimaldi A, Turpin G, Grosgogeat Y. Cardiac hypertrophy and function in asymptomatic acromegaly. Eur Heart J 1991;12:666-672.
52. Rodrigues EA, Caruana MP, Lahiri A, Nabarro JD, Jacobs HS, Raftery EB. Subclinical cardiac dysfunction in acromegaly evidence for a specific disease of heart muscle. Br Heart J 1989;62:185-194.
53. Colao A, Merola B, Ferone D, Lombardi G. Extensive personal experience: acromegaly. J Clin Endocrinol Metab 1997;82:2777-2781.
54. Minniti G, Jaffrain-Rea Ml, Moroni C, et al. Echocardiographic evidence for a direct effect of GH/IGF-I hypersecretion on cardiac mass and function in young acromegalics. Clin Endocrinol 1998;49:101-106.
55. Colao A, Baldelli R, Marzullo P, et al. Systemic hypertension and impared glucose tolerance are indipendently correlated to the severity of the acromegalic cardiomyopathy. J Clin Endocrinol Metab 2000;85:193.199.
56. Colao A, Cuocolo A, Marzullo P, et al. Impact of patient's age and disease duration on cardiac performance in acromegaly: a radionuclide angiography study. J Clin Endocrinol Metab 1999;84:1518-1523.
57. López-Velasco R, Escobar-Morreale HF, Vega B, et al. Cardiac involvement in acromegaly: specific myocardiopathy or consequence of systemic hypertension. J Clin Endocrinol Metab 1997;82:1047-1053.
58. Baldelli R, Ferretti E, Jaffrain-Rea ML, et al. Cardiac effects of slow-release lanreotide, a slow-release somatostatin analog, in acromegalic patients. J Clin Endocrinol Metab 1999;84:575-532.
59. Chanson P, Timsit J, Masquet C, et al. Cardiovascular effects of the somatostatin analog octreotide in acromegaly. Ann Intern Med 1990;113:921-925.
60. Merola B, Cittadini A, Colao A, et al. Chronic treatment with the somatostatin analog octreotide improves cardiac abnormalities in acromegaly. J Clin Endocrinol Metab 1993;77: 790-793.
61. Colao A, Cuocolo A, Marzullo P, et al. Effects of one-year treatment with octreotide on cardiac performance in patients with acromegaly. J Clin Endocrinol Metab 1999;84:17-23.
62. Colao A, Lombardi G. Growth hormone and prolactin excess. Lancet 1998;352:1455-1461.
63. Casanueva FF, Dieguez C. Growth hormone secretagogues: physiological role and clinical utility. Trends Endocrinol Metab 1999; 10:30-38.
64. Bowers CY, Sartor AA, Reynolds DG, Badger TAM. On the actions of the growth hormone-releasing hexapeptide, GHRP-6. Endocrinology 1991;128:2027-2035.
65. Howard AD. A receptor in pituitary and hypothalamus that functions in growth hormone release. Science 1996;273:974-977.
66. Rossoni G, De Gennaro Colonna V, Bernareggi M, Piovani GL, Muller EE, Berti F. Protectant activity of hexarelin or growth hormone against post-ischemic ventricular dysfunction in hearts from aged rats. J Cardiovasc. Pharmacol 1998;32:260-265.
67. Ong H, Bodart V, McNicoll N, Lamontagne D, Bouchard JF. Binding sites for growth hormone releasing peptide. GH and IGF Res 1998;8(Suppl B):137-140.
68. Tivesten A, Bollano E, Kujacic V, et al. The growth hormone secretagogue hexarelin improves cardiac function in rats after experimental myocardial infarction. Endocrinology 2000;141:60-66.

69. De Gennaro Colonna VG, Rossini M, Bernareggi M, Muller EE, Berti F. Cardiac ischemia and impairment of vascular endothelium function in hearts from growth hormone deficient rats. Protection by hexarelin. Eur J Pharmacol 1997;334:201-207.
70. Bisi G, Podio V, Valetto MR, et al. Acute cardiovascular and hormonal effects of GH and hexarelin, a synthetic GH-releasing peptide, in humans. J Endocrinol Invest 1999;22:266-272.

3
THE HEART AND THE GROWTH HORMONE DEFICIENCY SYNDROME

Mary Lee Vance
University of Virginia, Charlottesville, Virginia - USA

INTRODUCTION

In addition to promoting linear growth in children, growth hormone (GH) has a multitude of effects on numerous organ systems in children and adults. Over the past 20 years it has become apparent that GH is an important hormone for GH deficient (GHD) adults. The growth hormone deficiency syndrome includes abnormalities in body muscle, body water and adipose tissue, bone density, serum lipids, muscle strength, exercise tolerance and standardized measures of quality of life. Growth hormone replacement has beneficial effects on body composition, bone density, muscle strength, exercise endurance, serum lipids and quality of life measures. Specific consequences of GH deficiency on cardiac structure and function are an important aspect of GHD and are discussed below.

GHD IN ADULTS

Consequences of GHD: Three epidemiologic studies have confirmed that hypopituitary adults, not receiving GH replacement, have increased cardiovascular mortality compared with age and gender-matched subjects. Rosen and Bengtsson first described the cause of premature mortality in 333 adults with hypopituitarism who received standard hormone replacements; GH replacement was not then a part of the hormone replacement regimen in adults. All 53 patients who underwent testing were GH deficient and presumably the other patients were GH deficient as well. Mortality from cardiovascular disease was increased 1.9 fold compared with an age and gender matched population. Cardiovascular mortality was increased 1.7 fold in men and 2.5 fold in women. Of note, 76% of men received testosterone replacement and only 50% of women under 50 years received estrogen replacement (1). Thus, it is reasonable to speculate that cardiovascular mortality may have been reduced in the men who did not receive testosterone replacement and increased in the women who did not receive estrogen replacement. Also of note was the observation that mortality from malignancy was reduced by one half in hypopituitary adults compared with the normal population. One interpretation of this observation is that premature cardiac mortality occurred before the age-associated development of malignancy. In a study of 172 adults with hypopituitarism, Bates and colleagues reported a 1.73 fold increase in overall mortality in patients compared with an age and gender matched subjects. Vascular disease was the most common cause of death and was 1.35 fold increased over expected; hypogonadal patients had a better prognosis than those who received gonadal steroid replacement (2).

Cerebrovascular and cardiovascular mortality were increased in another study of 344 patients with hypopituitarism; the standardized mortality ratio for cerebrovascular disease was 3.39 and was 1.41 for cardiovascular disease. There was no increase in mortality from malignancy (3). These three studies indicate that hypopituitary adults not receiving GH replacement have premature cardiovascular mortality. An important issue is whether or not GH replacement reduces this risk. Because GH replacement in adults has been administered for approximately 12 years, it is not yet possible to state with certainty that GH replacement reverses this risk, but preliminary studies suggest that GH replacement is beneficial regarding cardiac structure and function.

Body composition: An increase in adipose mass, decrease in lean body mass, total body water and bone density occur in adults with both childhood and adult-onset GHD. One year after discontinuation of GH in 8 patients ages 14 – 21 years, quadriceps muscle strength, size and fiber areas decreased 6%, 5% and 14%, respectively (4). Several studies have confirmed the reduction in lean body mass in GHD adults. Total body adipose mass is increased 6 to 8 kg in GHD adults (5,6). Additionally, an increased waist:hip ratio suggests that visceral fat mass is increased (7), which is a known risk factor for development of cardiovascular disease. CT studies have demonstrated that intraabdominal and subcutaneous adipose mass is increased in adults with GHD (8).

Lipids: Abnormal serum lipid concentrations contribute to cardiovascular risk and adults with GHD have more lipid abnormalities than the normal population. In one study total serum cholesterol was 16% higher, LDL cholesterol was 25% higher, triglycerides were 43% higher and HDL cholesterol was 34% lower than in normal subjects (9). In another study, abnormally low serum HDL cholesterol concentrations were present in 72% of adult-onset and 54% of adults with childhood-onset GHD (10). Similar findings were reported in 34 GHD adults with 19% having increased total cholesterol and 23% having increased LDL cholesterol concentrations; no differences in HLD cholesterol were observed compared with a normal population (11). The severity of GHD, as indicated by GH responses to arginine and GHRH stimulation, correlated with the severity of lipid abnormalities in 157 adults with hypopituitarism. Total cholesterol, LDL cholesterol and triglycerides were highest and HLD cholesterol levels were lowest in adults with the lowest GH response to a stimulation test (12). Lipid abnormalities in GHD adults certainly contribute to the increased risk of cardiovascular disease.

Vascular Structure and Function: An ultrasound study of carotid and femoral arteries of 34 asymptomatic adults with hypopituitarism demonstrated that carotid intima-media thickness was 12% greater in GHD patients compared with normal subjects. The prevalence of atheromatous plaques was also greater in patients, 65% compared with 41% in control subjects. Addditionally, 32% of arteries in patients contained plaques while 18% of arteries in normal subjects contained plaques. Older patient age correlated with these vascular abnormalities in both groups (11). In 10 asymptomatic adults with GHD, aortic distensibility was decreased 23% compared with normal subjects; total cholesterol was also significantly higher in the patients (13). Thus, GHD affects vascular structure and arterial compliance which likely contributes to higher cardiovascular risk.

Cardiac Structure and Function: Numerous studies in animals and humans have amply demonstrated the importance of GH and IGF-1 on myocardial development and function.

Clinically, both excessive GH production (acromegaly) and GH deficiency have adverse effects on the heart, with both groups of patients having increased cardiovascular mortality.
The importance of GH on cardiac anatomy and function has been demonstrated in young asymptomatic adults who were treated with GH during childhood and in whom GH was discontinued after epiphyseal closure or satisfactory height was attained. In 12 such young adults studied at least 3 years after GH was discontinued, cardiac structure and function were abnormal compared with normal subjects. Echocardiography demonstrated that left ventricular posterior wall thickness was 33% less in GHD adults. Ejection fraction was decreased 17% at rest and during exercise was reduced 29% compared with normal subjects (14). In 11 asymptomatic GHD adults, 21 to 33 years, who had discontinued GH therapy at least 3 years earlier, interventricular septal thickness was decreased 22%, posterior ventricular wall thickness was 33% lower, left ventricular mass index was 36% lower and ejection fraction was 14% lower than in normal subjects (15). Similar abnormalities were found in 7 young GHD adults, 25 to 27 years; interventricular septal thickness, posterior wall thickness, left ventricular mass index and fractional shortening were significantly reduced compared with age matched normal subjects (16). Although most studies of young adults with GHD found no alterations in blood pressure or heart rate, Cittadini and colleagues noted that exercise associated heart rate was lower in young GHD adults (21 to 33 years) than in normal subjects and systolic blood pressure at rest and with exercise was also lower. Of particular note was the significant impairment in exercise duration in GHD young adults, 7.1 minutes compared with 10.3 minutes in normal subjects (17).
Equilibrium radionuclide angiography was used to study 11 adults, 21-33 years, with childhood onset and 9 adults, 19-40 years, with adult onset GHD. Left ventricular systolic performance at rest and during exercise was decreased in both groups of GHD adults compared with normal subjects. Ejection fraction at rest in both groups of GHD adults was 55% and was 66% in normal subjects. During exercise, ejection fraction was 54% in childhood onset and was 53% in adult onset patients compared with 76% in normal subjects. Additionally, GHD patients had a significant reduction in exercise tolerance as indicated by lower values of peak workload, peak rate-pressure product and exercise duration (18).
Echocardiographic and doppler studies were conducted in 13 men and 13 women, 26-64 years, with hypopituitarism, a low to normal serum IGF-1 concentration and an abnormal GH stimulation test. Left ventricular ejection fraction and left ventricular mass index were normal at rest. However, 50% had a prolonged isovolumic relaxation time and 31% had both an increased isovolumic relaxation time and abnormal left ventricular filling. During exercise, the left ventricular mass and maximum cardiac work were inversely related to the severity of GHD as assessed by the serum IGF-1 concentration. Additionally, left ventricular diastolic dysfunction and ST changes with exercise were considered to be a frequent finding (19).

SUMMARY

These studies amply demonstrate that GHD in adults has distinct pathologic effects on cardiovascular mortality, serum lipids and cardiovascular structure and function. Numerous studies in young adults with GHD confirm that myocardial anatomy and function are abnormal, providing a reasonable explanation for the premature mortality from cardiovascular and cerebrovascular events in adults with GHD. As discussed in this book, the effects of GH replacement appear very promising in reversing these risk factors.

REFERENCES

1. Rosen T, Bengtsson BA. Premature mortality due to cardiovascular disease in hypopituitarism. Lancet 1990;336:285-288.
2. Bates AS, Hoff WV, Jones PJ, Clayton RN. The Effect of hypopituitarism on life expectancy. J Clin Endocrinol Metab 1986;81:1169-1172.
3. Bulow B, Hagmart L, Mikoczyt Z, Nordstrom CH, Erfurth EM. Increased cerebrovascular mortality in patients with hypopituitarism. J Clin Endocrinol Metab 1997;46:75-81.
4. Rutherford OM, Jones DA, Round JM, Preece MA. Changes in skeletal muscle after discontinuation of growth hormone treatment in young adults with hypopituitarism. Acta Paediatr Scand 1989;356:61-63.
5. Binnerts A, Deurenberg P, Swart GR, Wilson JHP, Lamberts SWJ. Body composition in growth hormone deficient adults. Am J Clin Nutr 1992;55:918-923.
6. Rosen T, Bosaeus I, Tölli J, Lindstedt G, Bengtsson BA. Increased body fat mass and decreased extracellular fluid volume in adults with growth hormone deficiency. Clin Endocrinol 1993; 38:63-71.
7. De Boer H, Blok GJ, Van Der Veen EA. Clinical aspects of growth hormone deficiency in adults. Endocr Rev 1995;16:63-86.
8. Bengtsson BA, Eden S, Lönn L, et al. Treatment of adults with growth hormone (GH) deficiency with recombinant human GH. J Clin Endocrinol Metab 1993;76:309-17.
9. Cuneo RC, Salomon F, Watts GF, Hesp R, Sönksen PH. Growth hormone treatment improves serum lipids and lipoproteins in adults with growth hormone deficiency. Metabolism 1993;42:1519-1523.
10. Attanasio AF, Lamberts SWJ, Matranga AMC, et al. Adult growth hormone (GH)-deficient patients demonstrate heterogeneity between childhood onset and adult onset before and during human GH treatment. J Clin Endocrinol Metab 1997;82:82-88.
11. Markussis V, Beshyah SA, Fisher C, Sharp P, Nicolaides AN, Johnston DG. Detection of premature atherosclerosis by high-resolution ultrasonography in symptom-free hypopituitary adults. Lancet 1992;340:1188-1192.
12. Colao A, Berbone G, Pivonello R, et al. The growth hormone (GH) response to the arginine plus gh-releasing hormone test is correlated to the severity of lipid profile abnormalities in adult patients with gh deficiency. J Clin Endocrinol Metab 1999;84:1277-1282.
13. Lehmann ED, Hopkins KD, Weissberger AJ, Gosling RG, Sonksen PH. Aortic distensibility in growth hormone deficient adults. Lancet 1993;341:309.
14. Sacca L, Cittadini A, Fazio S. Growth Hormone and the Heart. Endocr Review 1994;15:555-573.
15. Merola B, Cittadini A, Colao A, et al. Cardiac structural and functional abnormalities in adult patients with growth hormone deficiency. J Clin Endocrinol Metab 1993;77:1658-1661.
16. Amato G, Carella C, Fazio S, et al. Body composition, bone metabolism, and heart structure and function in growth hormone (GH)-deficient adults before and after GH replacement therapy at low doses. J Clin Endocrinol Metab 1993;77:1671-1676.
17. Cittadini A, Cuocolo A, Merola B, et al. Impaired cardiac performance in GH-deficient adults and its improvement after GH replacement. Am J Physiol 1994;267:E219-E225.

18. Longobardi S, Cuocolo A, Merola B, et al. Left ventricular function in young adults with childhood and adulthood onset growth hormone deficiency. Clin Endocrinol 1998;48:137-143.
19. Shahi M, Beshyah SA, Hackett D, Sharp PS, Johnston DG, Foale RA. Myocardial dysfunction in treated adult hypopituitarism: a possible explanation for increased cardiovascular mortality. Br Heart J 1992;67:92-96.

4

GROWTH HORMONE REPLACEMENT AND CARDIAC FUNCTION IN THE GROWTH HORMONE DEFICIENT ADULTS

Leif Thuesen and Jens S. Christiansen
Department of Endocrinology M, Aarhus University Hospital
Aarhus Kommunehospitalet, Aarhus, Denmark

INTRODUCTION

The production of biosynthetic human growth hormone has opened the possibility for growth hormone (GH) substitution of adult growth hormone deficient (GHD) patients, and this treatment has been shown to improve muscle strength, exercise capacity and psychosocial well being in these patients (1,2). GH exerts significant effects on cardiac function and structure and both GH insufficiency and excess seem to be associated with increased cardiovascular morbidity and mortality (3,4). Therefore, it is of major clinical interest to examine the short- and long-term effect on the heart of GH substitution. The present chapter will review the literature on the cardiac effects of GH in hearts without prior GH deficiency, cardiac function in GHD patients and the cardiac effects of GH treatment in GHD patients. Here the latest long-term follow-up data of the Aarhus study will be presented.

EFFECTS OF GH IN HEARTS WITHOUT PRIOR GH DEFICIENCY

Subcutaneous GH in a dose of 9,6 IU/day for one week induced a reversible increase of heart rate, fractional shortening (left ventricular systolic function) and cardiac output without changing the thickness of the left ventricular wall (5). Similar, albeit more pronounced, alterations of cardiac function have been described in patients with short-term active acromegaly (6) and in animal experiments (7). Thus, GH excess for a shorter period of time seems to be associated with a condition of peripheral hyperperfusion and cardiac hyperkinesia. A such hyperkinetic syndrome might have deleterious effects on the cardiovascular system, if it persists for a longer period of time.

CARDIAC STRUCTURE AND FUNCTION IN GHD ADULTS

Cardiac structure and function in GHD patients has been compared with age- and sex-matched normal control subjects in at least five studies (8-12). The results are summarised in table 1. The prevailing findings were normal or decreased left ventricular wall-mass or -thickness, normal or decreased left ventricular systolic and diastolic function and normal or decreased cardiac output. Thus, GH deficiency seem to be associated with a subclinical impairment of the cardiac function, which at least in part might be responsible for the low exercise capacity in these patients (1).

Table 1. Cardiac function in GHD patients as compared to age- and sex-matched normal control subjects.

	GHD patients #	Left ventricular systolic function/ /cardiac output	Left ventricular diastolic function	Left ventricular wall thickness/ /mass
Amato et al. (8)	7	Decreased	-	Decreased
Merola et al. (9)	11	Decreased	-	Decreased
Thuesen et al. (10)	21	Normal	Normal	Normal
Valcavi et al. (11)	20	Normal	Decreased	Normal
Cuocolo et al. (12)	14	Decreased	-	-

CARDIAC FUNCTION AND STRUCTURE FOLLOWING GH SUBSTITUTION FOR 4-12 MONTHS

Cardiac function and structure following GH substitution have been investigated by echocardiography or nuclear angiography in at least 6 placebo controlled studies (8,10-12). Results are shown in table 2. The results are consistent and indicate, that GH substitution increases and probably normalises cardiac output, left ventricular systolic function, diastolic function and wall mass or wall thickness in GHD patients.

Table 2. Cardiac function in GHD patients following GH substitution therapy for 4-12 months as compared to placebo treatment.

	GHD patients #	Left ventricular systolic function/ /cardiac output	Left ventricular diastolic function	Left ventricular wall thickness/ /mass
Amato et al. (8)	7	Increased	-	Increased
Cittadini et al. (13)	11	Increased	-	Decreased
Thuesen et al. (10)	21	Increased	Normal	Normal
Valcavi et al. (11)	20	Increased	Increased	Increased
Cuocolo et al. (12)	14	Increased	-	-
Caidahl et al. (14)	10	Increased	-	-

CARDIAC FUNCTION AND STRUCTURE FOLLOWING LONG-TERM GH SUBSTITUTION

We performed Doppler echocardiography in 21 GH deficient patients after 4 months placebo and 4 months GH therapy, in a double blind cross-over study, the so call Aarhus Study (1). Some of the patients continued GH treatment in an open design study. Thus 13, 9, 8 and 5 of the patients were reinvestigated following GH treatment for 16, 38, 60 and 84 months, respectively. The long-term GH dose was 2 IU/m^2 Twenty-one age and sex-matched normal control subjects were also investigated. In table 3 the results are given for all the patients in the study.

Table 3. Heart rate (HR), systolic blood pressure (SBP), diastolic blood pressure (DBP). Left ventricular diastolic diameter (LVDD), left ventricular fractional shortening (FS), left ventricular wall mass (wall mass) and cardiac index (CI) in 5 GHD patients during 7 years of GH substitution therapy.

	Controls placebo		Growth Hormone substitution (months)				
			4	16	36	60	84
N	21	21	21	13	9	8	5
HR (beats/min)	63±9	69±11*	76±14**#	74±10	74±10	67±12	65±9
SBP (mmHg)	116±11	101±8**	105±11**	108±11	107±12	108±10	112±13
DBP (mmHg)	72±12	69±9**	70±10**	71±8	72±6	76±7	80±7
LVDD (mm)	50±4	46±6*	47±6*#	48±5	48±5	50±6	50±5
FS (%)	35±4	34±6*	35±6	40±6#	38±4#	37±9	40±9
E/A	1.7±0.3	1.9±0.5	1.8±0.5	1.8±0.4	1.9±0.5	1.7±0.6	1.9±0.5
Wall mass (g/m^2)	120±22	119±31	122±33	122±20	117±19	103±17	101±23
CI (L/min m^2)	2.5±0.5	2.5±0.8	3±1**#	3.1±0.8	3.1±0.4#	2.7±0.5	3.1±0.8

*p<0.05; **p<0.01 as compared to normal control subjects; #p<0.05 as compared to placebo treatment.

We found the heart rate to be increased in placebo treated patients as compared to controls. After 4 months of GH treatment, heart rate showed a further increase (10%, $P<0.01$) and seemed to remain elevated after 16 months of GH therapy. Systolic and diastolic blood pressures were significantly lower in placebo treated patients than in controls, and did not change significantly after GH treatment. The left ventricular diastolic diameter was reduced in patients as compared to controls, but increased after 4 months GH therapy ($P>0.05$) and seemed to increase further during 16 months of GH treatment. Cardiac index was at the same level in controls and in placebo-treated patients, but increased by 20% following GH therapy and remained elevated after 16 and 38 months ($P<0.05$). At the investigations after 60 and 84 months of GH substitution therapy no further changes were seen and there were no signs of adverse cardiovascular effects.

CONCLUSION

Short-term GH substitution in GHD adult patients seems to normalise the cardiovascular function. The long-term results are still based on a very limited number of patients, but in these patients there are no signs of deleterious cardiovascular effects of the GH substitution.

REFERENCES

1. Jørgensen JOL, Pedersen SA, Thuesen L, et al. Effects of growth hormone therapy to GH deficient adults on body composition, muscle strength, exercise capacity and cardiac and renal function. A double-blind crossover study. Lancet 1989;I:1221-5.
2. Salomon F, Cuneo RC, Hesp R, Sönksen PH. The effects of treatment with recombinant human growth hormone on body composition and metabolism in adults with growth hormone deficiency. N Eng J Med 1989; 321:1797-1803.

3. Rosen T, Bengtsson BA. Premature mortality due to cardiovascular disease in hypopituitarism. Lancet 1990;336:285-288.
4. Wright AD, Hill DM, Lowy C, Fraser TR. Mortality in acromegaly. Q J Med 1970;39:1-16.
5. Thuesen L, Christiansen JS, Sørensen KE, Ørskov H, Henningsen P. Increased myocardial contractility following growth hormone administration in normal man. An echocardiographic study. Dan Med Bul 1988;35:193-6.
6. Thuesen L, Christensen SE, Weeke J, Ørskov H, Henningsen P. A hyperkinetic heart in uncomplicated active acromegaly. Explanation for the high prevalence of hypertension in acromegalic patients. Acta Med Scand 1988;223:337-43.
7. Penney DG, Dunbar JC Jr, Baylerian MS. Cardiomegaly and haemodynamics in rats with a transplantable growth hormone-secreting tumour. Cardiovasc Res 1985;19:270-277.
8. Amato G, Carella C, Fazio S, et al. Body composition, bone metabolism, and heart structure and function in growth hormone (GH)-deficient adults before and after GH replacement therapy at low doses. J Clin Endocrinol Metab 1993;77:1671-1676.
9. Merola B, Cittadini A, Colao A, et al. Cardiac structural and functional abnormalities in adult patients with growth hormone deficiency. J Clin Endocrinol Metab 1993;77:1658-1661.
10. Thuesen L, Jorgensen JO, Muller JR, et al. Short and long-term cardiovascular effects of growth hormone therapy in growth hormone deficient adults. Clin Endocrinol (Oxf) 1994;41:615-620.
11. Valcavi R, Gaddi O, Zini M, Lavicoli M, Mellino U, Portioli I. Cardiac performance and mass in adults with hypopituitarism: effects of one year of growth hormone treatment. J Clin Endocrinol Metab 1995;80:659-666.
12. Cuocolo A, Nicolai E, Colao A, et al. Improved left ventricular function after growth hormone replacement in patients with hypopituitarism: assessment with radionuclide angiography. Eur J Nucl Med 1996;23:390-394.
13. Cittadini A, Cuocolo A, Merola B, et al. Impaired cardiac performance in GH-deficient adults and its improvement after GH replacement. Am J Physiol 1994;267:E219-E225.
14. Caidahl K, Eden S, Bengtsson BA. Cardiovascular and renal effects of growth hormone. Clin Endocrinol (Oxf) 1994;40:393-400.

5
THE HEART IN ACROMEGALY

Marco Losa and Klaus Von Werder*
Dept. of Neurosurgery, IRCCS San Raffaele, University of Milano, Italy
**Schlosspark Klinik, Humboldt University, Berlin, Germany*

INTRODUCTION

The existence of a specific cardiomyopathy due to growth hormone (GH)-hypersecretion has long been disputed mostly because the majority of patients included in clinical studies presented with other cardiovascular and metabolic alterations typically observed in acromegaly, such as hypertension and diabetes mellitus, which could impair cardiac function independently of GH hypersecretion. The prevalent opinion today, however, is to consider acromegalic cardiomyopathy as a specific entity characterized by cardiomegaly, cardiac hypertrophy, shortening of the ejection fraction of the left ventricle, fibrosis and degeneration of myocytes (1, 2). The existence of acromegalic patients affected by cardiomyopathy but without coexistent hypertension or diabetes mellitus strongly supports the idea of an independent detrimental effect on cardiac function due to chronic exposure to excessive GH levels (3, 4).

PATHOLOGICAL ABNORMALITIES OF THE HEART IN ACROMEGALY

Cardiomegaly and hypertrophy

The association between gross pathological changes of heart morphology and acromegaly was recognized more than a century ago when the disease was first described (5). Enlargement and hypertrophy have been considered the hallmarks of chronic exposure to excessive GH levels in the heart. Indeed some autopsy reports of acromegalic patients described the presence of hearts weighing more than 1000 g (6, 7). An autoptic study (8), performed on 27 acromegalic patients examined during a period of 60 years at the Mayo Clinic, reported on incidence of cardiomegaly up to 81%. Mean age at death was 50 years. Hypertension was present in 12 patients: interestingly 11 of the 12 hypertensive patients had cardiomegaly (91%). Nevertheless, cardiomegaly was detected in 11 of the 15 patients without hypertension (73%), underscoring the fact that hypertension favors but is not essential for the development of cardiomegaly in acromegalic patients (8). In this study, patients were subdivided according to the estimated duration of disease in group 1 (disease lasting less than 10 years) and group 2 (disease lasting 10 years or more): cardiomegaly was more frequently found in the 16 patients of group 1. Moreover the mean heart weight was higher in group 2 than in group 1 patients (618 g vs. 348 g, respectively). These data point out to the importance of duration of disease in determining the development of cardiomegaly. However, this association might have also been influenced by the older age and higher frequency of hypertension in group 2 patients. Cavity dilatation occurs only in the later stages of acromegaly and is secondary to congestive cardiac failure.

Hypertrophy of the heart, usually involving both ventricles, is also very frequently found in acromegalic patients. In the above mentioned autoptic series from the Mayo Clinic (8), 25 of the 27 patients (93%) had evidence of myocardial hypertrophy. However, the relationship between cardiac hypertrophy and duration of acromegaly in this series was not evident, since 91% and 94% of patients with acromegaly of less or more than 10 year duration had hypertrophy (8). Echocardiography has later confirmed the high incidence of cardiac hypertrophy in acromegalic patients. Several studies, that used stringent and accepted criteria to correlate the measured left ventricular mass with the body surface area, found cardiac hypertrophy in most patients with uncomplicated disease (4, 9, 10). Obviously, patients with coexisting hypertension tend to have more severe hypertrophy, as demonstrated by the significantly greater left ventricle mass in hypertensive compared to normotensive acromegalic patients (3, 10). The degree of cardiac hypertrophy could be correlated with the duration of disease but not with the height of GH and insulin-like growth factor-I (IGF-I) levels (3, 11, 12).The first step of cardiac disturbances in acromegaly is concentric hypertrophy involving both ventricles leading to a mass/volume ratio always higher than 1 (13). As disease progresses, hypertrophy of heart may become eccentric and the mass/volume ratio becomes lower than 1.

Coronary arteries

Data on the prevalence of coronary artery involvement are rather limited and heterogeneous, mostly because of the different methods of used to evaluate coronary artery diseases (1). The above mentioned autoptic study by Lie and coworkers (8) found that large coronary arteries were normal in the majority of the patients examined (89%). However, in the same study a significant wall thickening of the small intramural vessels was detected in 22% of cases (8), suggesting that involvement of the small rather than the large arteries may be another specific characteristic of acromegalic heart disease. The latter is of importance in deciding therapeutic options when dealing with angina pectoris in acromegalic heart disease.

Histological changes

Interstitial fibrosis represents the hallmark of cardiac hypertrophy in acromegalic patients. It was found in 85% of autopsied cases with myocardial hypertrophy (8). The collagen component is abundant and accounts, at least in part, for the increased ventricular mass of the acromegalic heart (7, 8). In a recent study performed on ventricular endomyocardial biopsy samples obtained from 10 patients with active acromegaly and cardiac dysfunction, varying from New York Heart Association class II to IV, the amount of interstitial collagen in combination with small foci of fibrotic scars was 8-fold greater than in control patients with mitral stenosis (14). Moreover, the degree of myocardial fibrosis was positively correlated with the estimated duration of disease (14). A histological picture resembling myocarditis, characterized by the presence of lymphomononuclear infiltration and small areas of monocyte necrosis, was detected in more than half of the cases examined at autopsy (8). The presence of interstitial fibrosis probably interferes with intracardial conduction and might explain the high prevalence of ventricular arrhythmias in acromegalic patients (15, 16). Supporting this hypothesis, resolution of acromegaly does not change the frequency of ventricular arrhythmias because myocardial fibrosis can not regress completely (15, 16).

PATHOPHYSIOLOGY

The alterations of cardiac morphology and function in acromegalic patients may result from both direct and indirect mechanisms. Among the latter, the main factors are hypertension, increased total body water content and plasma and red cell volume, diabetes mellitus and hyperinsulinemia. However, the presence of heart hypertrophy in selected acromegalic patients that do not have major coexisting risk factors for cardiovascular disease suggests the existence of a direct action of GH on the heart. Indeed, rat myocardial cells possess functionally active rGH receptors, whose density is even greater than that found in many other tissues (17), and administration of rGH to rats induces the formation of new contractile fibrils (18). Rat exposed chronically to elevated GH levels show an increase of enzymatic myocardial activity, which translates into a better performance (19). However, administration of pharmacological doses of GH to rats caused an increased intramyocardial accumulation of collagen (20). The latter abnormality is of special interest, since myocardial fibrosis seems to play a specific and important role in acromegalic cardiomyopathy. Besides to a direct action, GH is likely to influence heart morphology through local stimulation of IGF-I production (21), which increases the size of cultured cardiomyocytes and can induce the expression of muscle-specific genes, such as myosin light chain-2, troponin I, and the skeletal a-myosin isoform, in cultured rat myocytes (22). The development of acromegalic cardiomyopathy can be divided in three phases, which are characterized by different morphological and functional events (1). It must be stressed that in untreated patients there is a gradual progression from one phase to the successive stage that is accelerated by hypertension and diabetes mellitus.

Early stage (hyperkinetic phase)

The first stage of acromegalic heart disease is represented by an increase of heart rate and cardiac output, whereas the peripheral resistance is usually diminished (23). This is probably due to the positive inotropic effect of GH on heart, which has been well demonstrated in normal subjects who received exogenous GH for a short period (24). IGF-I seems to play also a role in this early phase of cardiac involvement. Indeed, administration of IGF-I to rats induces morphological and physiological effects that lead to a clear enhancement of cardiac contractility (19, 25). Apart from the direct positive inotropic effect of GH and IGF-I on the heart, the other factor mediating the initial hyperkinetic response seems to be the increase of plasma volume due to the well-known sodium-retaining effect of GH and the inhibition of atrial natriuretic factor release. The increase of end-diastolic pressure (preload) does not occur, because of the adaptation of the peripheral vascular system, thus explaining the absence of cavity dilation in acromegalic cardiomyopathy not accompanied by heart failure. The early stage of acromegalic cardiomyopathy, therefore, can be considered as almost purely functional. No morphological changes of heart structure can be detected at this early stage.

Intermediate stage

As acromegaly progresses, evidence of diminished cardiac performance becomes apparent replacing the improved functional characteristics of the hyperkinetic phase. The positive inotropic action of GH is counterbalanced by the other negative actions that become manifest during chronic exposure to elevated levels of both GH and IGF-I levels. The intermediate phase of acromegalic cardiomyopathy is characterized by the detection of morphological abnormalities, that are not present in the hyperkinetic phase. The progressive accumulation of collagen that leads to interstitial fibrosis seems to be of particular importance. Detection of increased myocyte and interstitial cells apoptosis in acromegalic hearts (14) could explain the initiation of the hypertrophic response. Although apoptosis occurs without collagen accumulation in the involved organs, loss of myocyte cell mass should impair the ability of the myocardium to develop force, thus decreasing pump function and ejection fraction (26). Progressive loss of myocardial cells negatively influences anatomy and hemodynamics of the heart. Physiological alterations of cardiac function are accompanied by restructuring of the wall with side-to-side slippage of myocytes and mural thinning. At this point interstitial fibrosis may result from the reaction of connective tissue cells to pathological loads (27), thus initiating a vicious cycle leading to progressive cardiac dysfunction and cardiac enlargement. The latter is usually due to pronounced concentric myocardial hypertrophy, as shown by several echocardiographic studies (3, 4, 12, 15, 28, 29). Noteworthy, progressive collagen accumulation in rats initially leads to reduced diastolic function, whereas systolic dysfunction appears later when myocardial fibrosis becomes more extensive (30). The time course of acromegalic cardiomyopathy strictly resembles this experimental model. Indeed, the first functional abnormality in the intermediate phase of acromegalic cardiomyopathy involves diastolic function at rest. Studies using Doppler echocardiography demonstrated that abnormal diastolic filling of both the right and left ventricles was a characteristic finding in acromegaly (3, 4). The principal alterations included a prolonged relaxation time, a decrease of the ratio of early to late mitral and tricuspid flow velocities, decreased diastolic filling wave and increased reversal flow during atrial contraction (1). The increased stiffness of the ventricles is explained by the combined effects of hypertrophy and interstitial fibrosis. These abnormalities were present both in normotensive and hypertensive patients, even though the latter usually showed more marked alterations. Definitive proof that diastolic dysfunction is not caused by coexisting hypertension was provided by a recent study in a group of 11 normotensive acromegalic patients who did not have evidence of overt cardiac disease. Doppler echocardiography showed the same alterations described in unselected acromegalic patients and, again, measurements reflecting the diastolic function of the left ventricle were particularly abnormal when compared to normal controls (31). Since assessment of diastolic and systolic functions by Doppler echocardiography may be limited by the intra- and interobserver variability and poor sensitivity of the method, it is important that studies using a different methodology, namely radionuclide angiography, resulted into similar findings (10, 15, 32, 33). Colao and coworkers (33) studied a large group of acromegalic patients without additional complications affecting heart function and normal controls subdivided according to age: they demonstrated that left ventricular filling was especially impaired in acromegalic patients older than 40 years, suggesting that both patient's age and estimated duration of disease are relevant factors for the development of diastolic dysfunction. While systolic function at rest is still not compromised, the other functional abnormality, appearing at this stage of acromegalic cardiomyopathy, is impaired performance on effort. In a study using radionuclide angiography, a group of acromegalic patients without overt heart disease and hypertension showed indices of systolic function at rest similar to that of a control group, but during exercise the ejection fraction did not increase, as it would be expected, in the acromegalic group (10). Colao and coworkers (33) suggested that patient's age and estimated duration of disease were the most important factors in determining the degree of functional impairment during exercise. Indeed, 61% of acromegalic patients younger than 40 years showed the normal increase (> 5% of basal value) of the ejection fraction at peak exercise, whereas only 4% of the patients older than 40 years had the normal response to exercise (33). The principal and often only clinical symptom due to impairment of cardiac function in the intermediate stage of the disease is dyspnea during exertion (10).

End stage

If acromegaly progresses unabated, heart involvement becomes clinically more important leading to increased morbidity and mortality. The morphological abnormalities typical of the intermediate phase of the disease worsen gradually. Cardiac hypertrophy reaches unusual proportions and interstitial and myocardial fibrosis advance to the point that contractility decrease. However, the principal morphological characteristic of this phase is ventricular dilatation, which, together with extensive biventricular hypertrophy, leads to the impressive degree of cardiomegaly found in some acromegalic patients. From a clinical point of view, cardiac performance, which was impaired only on effort, begins to deteriorate even at rest. Catheterization studies have shown that the ventricular end-diastolic pressure increases and the ejection fraction decreases, resulting in the final picture of overt congestive heart failure (34).

EPIDEMIOLOGICAL STUDIES

Cardiovascular diseases

Hypertension is a frequent complication of acromegaly. Its prevalence varied between 18% and 51% in the different series reported (2, 11, 15, 35-42; Table 1). Compared to age- and sex-matched controls, acromegalic patients showed a 3-fold risk of hypertension (39).

Table 1. Prevalence of hypertension in acromegalic patients.

Author	No. of patients	Prevalence (%)
Gordon (1962)	100	18
McGuffin (1974)	57	23
Martins (1977)	16	25
Mather (1979)	23	35
Smallridge (1979)	27	22
Jadresic (1982)	155	32
Csanády (1983)	25	40
Nabarro (1987)	256	31
Rodrigues (1989)	34	26
Kraatz (1990)	158	31
Fazio (1993)	30	30
Ezzat (1994)	500	51
Ohtsuka (1995)	64	38
Lopez-Velasco(1997)	39	33

It can not be ruled out that the reported variability in the frequency of hypertension in acromegaly may be due to the difference in investigating and defining arterial hypertension. Some series did not find any sex-related difference in the incidence of hypertension (39, 41), whereas other investigators (40) detected a higher frequency in female patients. It has been suggested that the frequency of hypertension is higher with increasing age of patients and a longer estimated duration of disease (11). There seems to be no correlation between basal levels of either GH or IGF-I and the frequency or the severity of hypertension. Recently, it has been suggested that a positive familiarity may increase the risk of hypertension in acromegalic patients, possibly by enhancing IGF-I production (42). Usually hypertension in acromegaly is not very severe and can be controlled successfully with one or more antihypertensive drugs. However, hypertension is an important prognostic factor in acromegaly. As mentioned above, coexisting hypertension is not necessary for the development of cardiomyopathy, but its presence significantly accelerates the course of disease. Moreover, hypertension was found to be an independent risk factor of mortality in acromegalic patients (43). Effective treatment of acromegaly, either by transsphenoidal surgery (44) or medical therapy with somatostatin analogs (45) and bromocriptine (39), causes a significant lowering of blood pressure in most hypertensive acromegalic patients: in some cases antihypertensive medication can be stopped completely.

The prevalence of coronary artery disease in acromegaly has not been established conclusively, mainly because of the various definitions (clinical or instrumental) used to document this complication. In a large casistic, clear clinical and electrocardiographic evidence of angina pectoris or a documented myocardial infarction in life or at autopsy was found in 10.5% of acromegalic patients (39). In another series angina pectoris was present in 25% of patients (29). The prevalence of symptoms attributable to coronary artery disease was lower than 5% in other series (11, 28, 37). Studies based on myocardial perfusion scintigraphy demonstrated a higher frequency, ranging from 20% to 37%, of coronary artery disease (10, 15, 36). On the whole, the prevalence of clinically relevant coronary artery disease in acromegaly does not seem to be particularly high when compared to age- and sex-matched controls (39). However, the prognostic importance of this complication in acromegaly is underscored by the fact that only few patients experience improvement of ischemic symptoms after remission of acromegaly. The improvement of clinical symptoms attributable to ischemia seems to be secondary in most of these cases to amelioration of coexisting hypertension (39).

Arrhythmias are increasingly recognized as an important complication of acromegaly. A predisposing factor for the development of arrhythmia is the above mentioned increased myocardial fibrosis and enhanced collagen deposition with focal areas of fiber disarray (8). The latter would provide an anatomical substrate for a slowing and alteration of the electrical activity of the heart. Moreover, a prolongation of the duration of action potentials and of the depolarization, due to alterations of K^+ and Ca^{++} flux, has been described in an animal model with chronic GH excess (46). Holter monitoring discovered a very high (41%) frequency of ventricular extrasystolia in a group of 34 acromegalic patients without known cardiopathy of other cause (15). Moreover, more worrisome abnormalities were detected in other 4 patients: 2 cases with ventricular couplets, one case with symptomatic ventricular tachycardia and one case with asymptomatic sinus arrest (15). An increased prevalence of prognostically dangerous ventricular arrhythmias (Lown class III and IV) was also found in another study performed in 32 patients (47). In the latter study, the severity of arrhythmias did not correlate with basal GH levels, but it was associated with the estimated duration of the disease, with a derived clinical score of disease activity, and with the left ventricular mass (47). The reversibility of arrhythmias after remission of acromegaly is controversially discussed: in one study the frequency of ventricular extrasystolia did not change after elective treatment of acromegaly (47), whereas in another study 3-6 months of therapy with octreotide significantly improved several of the electrocardiographic abnormalities detected before therapy. Furthermore, ventricular and supraventricular tachicardia disappeared in all treated patients (45)

Cardiovascular mortality

Acromegaly has been associated with increased overall morbidity and mortality. Several epidemiological studies suggest that the mortality rate in acromegalic patients is approximately double that found in the general population (48-51). It has been reported that acromegalic patients have an estimated life-expectancy of 10 years lower than that of the general population (43, 48). Even though malignant and respiratory diseases are associated to a higher mortality risk, cardiovascular disease is the predominant cause of enhanced mortality in acromegaly, as demonstrated by several epidemiological studies (8, 39, 48-52; Table 2). The proportion of deaths due vascular causes (including cerebrovascular disease) accounts for approximately 40-60% of the total deaths in acromegalic patients (39, 48-52). In one study (43) the data on overall mortality were further analyzed by Cox's multivariate analysis and it was found that among several variables the major predictors of mortality were, in decreasing order of importance, the last known GH level, diagnosis of cardiovascular disease at the time of diagnosis, presence of hypertension at diagnosis and duration of symptoms before diagnosis. Of particular importance is the relationship between posttreatment GH level and mortality. Bates and coworker (51) found that the higher the levels of GH, both before and after treatment, the higher the mortality rate. However, when mean GH levels are less than 2.5 μg/L, independently of the treatment modality, the mortality rate was not different from that of the general population. This finding has been often regarded to indicate that the aim of treatment of acromegalic patients should be to reach a "safe" GH level and that patients with otherwise other signs of active disease (e.g., incomplete

Table 2. Observed to expected number of deaths and standardized mortality rates with 95% confidence intervals due to cardiovascular disease in acromegalic patients. Deaths due to cerebrovascular disease were not counted.

Author and No. of patients	Observed to expected deaths	Standardized mortality rates	95% Confidence intervals
Alexander (1980) (n=164)	13 / 4.5	2.89	NR
Nabarro (1987) (n=256)	19 / 12.7	1.50	NR
Rajasoorya (1994) (n=151)	15 / 5	3.00	NR
Orme (1998) (n=1362)	134 / 76	1.76	1.47 - 2.07

NR = Not Reported

GH suppression during glucose load, persistence of a paradoxical response to TRH/GnRH, or slight elevation of IGF-I) have, nevertheless, a good long term prognosis as long as "safe" GH levels are maintained. Another study confirmed that acromegalic patients with posttreatment levels below 2.5 μg/L have an overall mortality rate comparable to the general population. Moreover, the same relationship was found also when deaths due to cardiovascular disease were analyzed separately (52). In particular the standardized mortality rates for death due to cardiovascular disease were 1.20, 1.59, and 2.11 in patients with posttreatment GH level < 2.5 μg/L, comprised between 2.5 μg/L and 9.9 μg/L, and □ 10 μg/L, respectively (52). The relationships between duration of acromegaly and age at diagnosis with cardiovascular mortality were not significant, at variance with the near significant ($p = 0.058$) or significant ($p = 0.004$) association of both variables with overall mortality (52). Of particular concern in analyzing the data on cardiovascular mortality is the potential confounding effect of coexistent hypopituitarism, as the latter has also been associated to an increased risk of cardiovascular mortality (53). Bates and coworkers (51) did not find any excess of pituitary hypofunction in the deceased subgroup of patients, but the other studies did not apparently control for this factor.

SUMMARY

Recognition and appreciation of the specific and sometimes severe alterations (Table 3) induced by excess GH secretion in acromegalic patients is a rather recent event. It has been especially difficult to ascribe the impairment of cardiac function to the direct effects of GH and/or IGF-I on heart rather than to coexisting hypertension and glucose intolerance, which often complicate the course of acromegalic disease.

Elucidation of the direct effects at the cellular level exerted by GH and IGF-I on the heart and several clinical studies performed in acromegalic patients have led most authors to regard acromegalic cardiomyopathy as a specific nosologic entity, characterized by different stages of progression. This point has also been accepted by the World Health Organization that has added acromegaly in the group of systemic disease causing a specific cardiomiopathy (54).

Table 3. Cardiac manifestations of acromegaly

• Ventricular hypertrophy
• Mitral valve insufficiency
• Interstitial fibrosis
• Myocarditis (lymphocytic infiltration)
• Ventricular arrhytmia
• Impaired diastolic filling
• Impaired exercise performance
• Congestive heart failure

REFERENCES

1. Saccà L, Cittadini A, Fazio S. Growth hormone and the heart. Endocr Rev 1994;15:555-573.
2. Lopez-Velasco R, Escobar-Morreale H, Vega B, et al. Cardiac involvement in acromegaly: specific myocardiopathy or consequence of systemic hypertension? J Clin Endocrinol Metab 1997;82:1047-1053.
3. Fazio S, Cittadini A, Sabatini D, et al. Evidence of biventricular involvement in acromegaly: a Doppler echocardiographic study. Eur Heart J 1993;14:26-33.
4. Morvan D, Komajda M, Grimaldi A, et al. Cardiac hypertrophy and function in asymptomatic acromegaly. Eur Heart J 1991;12:666-672.

5. Huchard H. Anatomie pathologique, lesions et troubles cardiovasculaires de l'acromegalie. J Practiciens 1895;9:249- 253.
6. Courville C, Mason VR. The heart in acromegaly. Arch Intern Med 1938;61:704- 710.
7. Hejtmancik MR, Bradfield JY, Hermann GR. Acromegaly and the heart: a clinical and pathologic study. Ann Intern Med 1951;34:1445-1446.
8. Lie JT, Grossman SJ. Pathology of the heart in acromegaly: anatomic findings in 27 autopsied patients. Am Heart J 1980;100:41-52.
9. Hradec J, Marek K, Kral J, Janota T, Poloniecki J, Malik M. Long-term echocardiographic follow-up of acromegalic heart disease. Am J Cardiol 1993;72:205-210.
10. Fazio S, Cittadini A, Cuocolo A, et al. Impaired cardiac performance is a distinct feature of uncomplicated acromegaly. J Clin Endocrinol Metab 1994;79:441-446.
11. McGuffin Jr WL, Sherman BM, Roth J, et al. Acromegaly and cardiovascular disorders. A prospective study. Ann Intern Med 1974;81:11-18.
12. Savage DD, Henry WL, Eastman RC, Borer JF, Gorden P. Echocardiographic assessment of cardiac anatomy and function in acromegalic patients. Am J Med 1979;57:823-829.
13. Lombardi G, Colao A, Marzullo P, et al. Is growth hormone bad for your heart? Cardiovascular impact of GH deficiency and acromegaly. J Endocrinol 1997;155:S33-S37.
14. Frustaci A, Chimenti C, Setoguchi M et al. Cell death in acromegalic cardiomyopathy. Circulation 1999;99:1426-1434.
15. Rodrigues EA, Caruana MP, Lahiri A, Nabarro JDN, Jacobs HS, Raftery EB. Subclinical cardiac dysfunction in acromegaly: evidence for a specific disease of heart muscle. Br Heart J 1989;62:185-194.
16. Kahaly G, Syover C, Beyer J, Mohr-Kahaly S. Relation of endocrine and cardiac findings in acromegalics. J Endocrinol Invest 1992;15:13-18.
17. Mathews LS, Enberg B, Norstedt G. Regulation of rat growth hormone receptor gene expression. J Biol Chem 1989;264:9905-9910.
18. De Grandpre R, Raab W. Interelated hormonal factors in cardiac hypertrophy, experiments in nonhyperthensive hypophysectomized rats. Circ Res 1953;1:345-353.
19. Timsit J, Riou B, Bertherat J, et al. Effects of chronic growth hormone hypersecretion on intrinsic contractility, energetics, isomyosin pattern, and myosin adenosine triphosphatase activity of rat left ventricle. J Clin Invest 1990;86:507-515.
20. Bruel A, Oxlund H. Biosynthetic growth hormone increases the collagen deposition rate in rat aorta and heart. Eur J Endocrinol 1995;132:195-199.
21. Isgaard J, Nilsson A, Vikmann K, Isaksson OGP. Growth hormone regulates the level of insulin-like growth factor-1 mRNA in rat skeletal muscle. J Endocrinol 1989;120:107-112.
22. Ito H, Hiroe M, Hirata Y, et al. Insulin like growth factor-I induces hypertrophy with enhanced expression of muscle specific genes in cultured rat cardiomyocytes. Circulation 1993;87:1715-1721.
23. Thuesen L, Christensen SE, Weeke J, Ørskov H, Henningsen P. A hyperkinetic heart in uncomplicated active acromegaly: explanation of hypertension in acromegalic patients? Acta Med Scand 1988;223:337-343.
24. Thuesen L, Christiansen JS, Sørensen KE, Jorgensen JOL, Orskov H, Henningsen P. Increased myocardial contractility following growth hormone administration in normal man. Dan Med Bull 1988;35:193-196.

25. Vetter U, Kupferschmid C, Lang D, Pentz S. Insulin-like growth factors and insulin increase the contractility of neonatal rat cardiocytes *in vitro*. Basic Res Cardiol 1988;83:647-654.
26. Cheng W, Li B, Kajstura J. Stretch-induced programmed myocite cell death. J Clin Invest 1995;96:2247-2259.
27. Weber KT, Brilla CG. Pathological hypertrophy and cardiac interstitium fibrosis and renin-angiotensin-aldosterone system. Circulation. 1991;83:1849-1865.
28. Smallridge RC, Rajfer S, Davia J, Schaaf M. Acromegaly and the heart. An echocardiographic study. Am J Med 1979;66:22-27.
29. Csanády M, Gáspár L, Hgye M, Gruber N. The heart in acromegaly: an echocardiographic study. Int J Cardiol 1983;2:349-357.
30. Jalil JE, Doering CW, Janicki JS, Pick R, Shroff SG, Weber KT. Fibrillar collagen and myocardial stiffness in the intact hypertrophied rat left ventricle. Circ Res 1989;64:1041-1050.
31. Özbey N, Öncul A, Bugra Z, et al. Acromegalic cardiomyopathy: Evaluation of the left ventricular diastolic function in the subclinical stage. J Endocrinol Invest 1997;20:305-311.
32. Sicolo N, Bui F, Sicolo M, et al. Acromegalic cardiopathy: a left ventricular scintigraphic study. J Endocrinol Invest 1993;16:123-127.
33. Colao A, Cuocolo A, Marzullo P, et al. Impact of patient's age and disease duration on cardiac performance in acromegaly: a radionuclide angiography study. J Clin Endocrinol Metab 1999;84:1518-1523.
34. Pepine CJ, Aloia J. Heart muscle in acromegaly. Am J Med 1970;48:530-534.
35. Gordon DA, Hill FM, Ezrin C. Acromegaly: a review of 100 cases. Can Med Ass 1962;87:1106-1109.
36. Martins JB, Kerber RE, Sherman BM, Marcus ML, Ehrhardt JC. Cardiac function in acromegaly. Circulation 1977;56:863-869.
37. Mather HM, Boyd MJ, Jenkins JS. Heart size and function in acromegaly. Br Heart J 1979;41:697-701.
38. Jadresic A, Banks LM, Child DF, et al. The acromegaly syndrome relation between clinical features, growth hormone values and radiological characteristics of the pituitary tumours. Q J Med 1982;51:189-204.
39. Nabarro JDN. Acromegaly. Clin Endocrinol (Oxf) 1987;26:481-512.
40. Kraatz C, Benker G, Weber F, Ludecke D, Hirche H, Reinwein D. Acromegaly and hypertension: prevalence and relationship to the renin-angiotensin-aldosterone system. Klin Wochenschr 1990;68:583-587.
41. Ezzat S, Forster MJ, Berchtold P, Redelmeier DA, Boerlin V, Harris AG. Acromegaly. Clinical and biochemical features in 500 patients. Medicine (Baltimore) 1994;73:233-240.
42. Ohtsuka H, Komiya I, Aizawa T, Yamada T. Hypertension in acromegaly: hereditary hypertensive factor produces hypertension by enhancing IGF-I production. Endocr J 1995;42:781-787.
43. Rajasoorya C, Holdaway IM, Wrightson P, Scott DG, Ibbertson HK. Determinants of clinical outcome and survival in acromegaly. Clin Endocrinol (Oxf) 1994;41:95-102.
44. Giovanelli M, Losa M, Mortini P. Acromegaly: surgical results and prognosis. In: Landolt AM, Vance ML, Reilly PL (eds): Pituitary adenomas. Churchill Livingstone, Edinburgh, 1996:333-351.

45. Colao A, Ferone D, Cappabianca P, et al. Effect of octreotide pretreatment on surgical outcome in acromegaly. J Clin Endocrinol Metab 1997;82:3308-3314.
46. Xu X, Best P. Decreased transient outward K^+ current in ventricular myocytes from acromegalic rats. Am J Physiol 1991;260:H935-H942.
47. Kahaly G, Olshausen KV, Kahaly-Mohr S, et al. Arrhythmia profile in acromegaly. Eur Heart J 1992;13:51-56.
48. Alexander L, Appleton D, Hall R, Ross WM, Wilkinson R. Epidemiology of acromegaly in the Newcastle region. Clin Endocrinol (Oxf) 1980;12:71-79.
49. Wright AD, Hill DM, Lowy C, Fraser TR. Mortality in acromegaly. Q J Med 1970;39:1-16.
50. Bengtsson B, Eden S, Ernest I, Oden A, Sjogren B. Epidemiology and long term survival in acromegaly. Acta Med Scand 1988;223:327-335.
51. Bates AS, Van't Hoff J, Jones JM, Clayton RN. An audit of outcome of treatment in acromegaly. Q J Med 1993;86:293-299.
52. Orme SM, McNally RJQ, Cartwright RA, Belchetz PE for the United Kingdom Acromegaly Study Group. Mortality and cancer incidence in acromegaly: a retrospective cohort study. J Clin Endocrinol Metab 1998;83:2730-2734.
53. Rosen T, Bengtsson B. Premature mortality due to cardiovascular disease in hypopituitarism. Lancet 1990;336:285-288.
54. Richardson P, McKenna WJ, Bristow M, et al. Report of the 1995 WHO/ISFC task force on the definition and classification of cardiomyopathies. Circulation 1996;93:841:847.

6
IMPACT OF GROWTH HORMONE-LOWERING TREATMENTS ON HEART FUNCTION IN ACROMEGALY

Philippe Chanson
University Hospital Bicêtre, Assistance Publique-Hôpitaux de Paris, and Faculté de médecine Paris-Sud, 94275 Kremlin-Bicêtre, France

INTRODUCTION

Acromegaly is associated with an increased risk of premature death (1-8). Cardiac complications are known to be major determinants of the shortened life expectancy which is about 10 years compared with that of the general population (4, 6). A recent study by Orme *et al.* (8) has confirmed previous published work (5, 6) by showing that cardiovascular mortality rate increased with the posttreatment GH levels. The cardiovascular mortality rate was similar to that found in the general population [standardized mortality ratio (SMR), 1.20 (CI, 0.83–1.68)] when GH levels were < 2.5 μg/L, but tended to increase significantly with higher posttreatment GH levels (8) (Table 1).The etiology of the cardiovascular disease in acromegaly is multifactorial, as reviewed elsewhere in this book. A specific cardiomyopathy is likely, due to direct effect of GH/IGF-1 excess on cardiomyocytes, responsible for abnormalities in both cardiac muscle structure and function (9-12). Associated cardiovascular risk factors such as hypertension, diabetes mellitus, sleep apnea syndrome may also contribute to worsen cardiac function.

Treatment of acromegaly, by reducing GH levels, allows a substantial decrease in morbidity and mortality. Reversibility of cardiovascular complications cannot accurately be assessed without a precise definition of cure for acromegaly. More stringent criteria of cure are now proposed (13). These criteria have been revised in the light of physiological studies determining more precisely the normal GH and IGF-1 levels in normal individuals. Epidemiological studies have also provided evidence for validating these new criteria by demonstrating that mortality (and particularly, cardiovascular mortality) in acromegalic patients tends to be similar to that of the general population when post-treatment GH levels are less than 2.5 μg/L (6, 8, 13-15) (Table 1).

However these new stringent criteria of cure have been used only recently. As many patients with persistenly active acromegaly, have wrongly been considered as cured in earlier studies evaluating the effects of treatment on cardiac function and structure, their results appear very heterogenous.Moreover previous studies addressing the question of reversibility of cardiomyopathy in treated acromegalic patients have been hampered by the absence of rapidly and uniformly effective therapeutic modality for lowering the GH level. Surgery, radiation therapy and bromocriptine have only been able to reduce growth hormone levels to normal levels in 10 to 30 % of patients (16).

Table 1. Relationship between postreatment GH levels and cardiovascular and overall mortality in acromegaly. SMR, standardized mortality ratio, was calculated from expected deaths in the general population and observed deaths in the acromegalic population (n = 1362). Adapted from Orme *et al.* (8).

Cause of death	SMR (95% CI) / Post-treatment GH level:			χ^2 test for linear trend (P)
	< 2.5 μg/l	2.5 - 9.9 μg/l	□ 10 μg/l	
Cardiovascular disease	1;20 (0.83-1.68)	1.59 (1.15-2.15)	2.11 (1.42-3.01)	0.02
All causes of death	1.10 (0.89-1.35)	1.41 (1.16-1.68)	2.12 (1.70-2.62)	<0.0001

These studies essentially went to the conclusion that GH normalization had no role for preventing or controlling cardiac disease in acromegaly. In fact, earlier studies evaluating the effects of treatment of acromegaly involved patients with severely impaired cardiac function, at a stage of congestive cardiac failure, while more recent studies assessing reversibility of cardiomyopathy were conducted on acromegalic patients with earlier stages of the disease, limited to hyperkinetic heart syndrome and cardiac hypertrophy. The availability of medical treatment with somatostatin analogues able to rapidly normalize GH/IGF-1 levels in the majority of patients have led to a reappraisal of this question by showing that patients with cardiomyopathy may not only improve cardiac function and structure but also reverse to normal cardiac morphology. On the other hand, even after normalization of GH/IGF-1 levels, a proportion of patients keep cardiac abnormalities, at a variable degree. The age of the patients and the duration of the disease are probably important determinants in the reversibility of acromegalic cardiomyopathy (12, 17, 18).

NATURAL HISTORY OF ACROMEGALIC CARDIOMYOPATHY.

Before discussing the role of GH-lowering treatments on acromegalic cardiomyopathy, it is necessary to review the spontaneous course of cardiac complications in the absence of effective treatment of acromegaly. Longitudinal observational studies in man are obviously unavailable or give incomplete informations. Thus natural history of cardiomyopathy in acromegaly has to be constructed from cross-sectional studies, conducted in different forms of acromegalic cardiomyopathy (9).

At the early stage, no demonstrable change in morphological examination are associated with increased contractility, decrease vascular resistance and high cardiac output (the hyperkinetic heart syndrome). At the intermediate stage (no precise information about the minimal time and the importance of GH hypersecretion required for this stage to be reached, are currently available), ventricular hypertrophy becomes evident and diastolic function slowly deteriorates. The patient is usually asymptomatic, until ventricular thickening and fibrosis increase leading to worsening of diastolic function and impairment of systolic performance on effort. The end-stage is a consequence of untreated or unsufficient treatment of acromegaly. Fortunately this stage is observed more rarely nowdays. Cardiac function at this stage further deteriorates and cardiac congestive failure may develop. In this stage, cardiac output is greatly reduced and filling pressures increase, due to impairment of systolic ventricular function. Ventricular dilatation may be observed. Very few studies have

had the opportunity to evaluate sequentially cardiac function in acromegalic patients. Anecdotal case reports are generally limited to patients with cardiac congestice failure whose diagnosis of acromegaly was made at a severe stage of the disease (19-21).
Whatever its stage the deterioration of cardiac function is slow. In a study performed in 1969, only patients with obvious cardiac disease slowly impaired cardiac function during the 5-6 yr follow-up (22). In a more recent study, on the basis of a 10-yr prospective follow-up of acromegalic patients with persistently active disease, a significant increase in mean posterior wall thickness, left ventricular mass and left atrial diameter was demonstrated but no patient was reported to develop congestive cardiac failure (23). In conclusion, if the early stages of the natural history of acromegalic cardiomyopathy with progessive diastolic dysfunction and myocardial hypertrophy seem well established, their precise time-course is unknown (probably between 5 and 10 years of disease duration). The age of the patients is probably crucial, cardiac hypertrophy (independently for the presence of hypertension) being more marked in younger patients (12, 17, 18).
By contrast, if all patients would develop cardiac failure in the absence of treatment of acromegaly is presently unknown. In this setting, the combination of acromegaly with other conditions deleterious for the heart (such a diabetes mellitus, coronary heart disease, hypertension) is probably important but is not mandatory. The role of interstitial fibrosis (9, 12, 19-21, 24-27) and apoptosis (28) in this process has to be emphasized. However their exact relationship with acromegalic hypertrophy (and/or GH/IGF-1 excess) remains to be established. In this setting, it is surprizing to observe that animal models of GH excess do not develop cardiac dysfunction while they demonstrate hypercontractility and cardiac hypertrophy (29, 30). This may be related to the short duration of GH exposure or to species characteristics. Conversely, follow-up of cardiac involvement in spontaneously acromegalic cats did demonstrate the frequent development of cardiac failure in these animals (31).

CARDIAC EFFECTS OF "TRADITIONAL TREATMENT MODALITIES" IN ACROMEGALY.

Attemps to reverse cardiomyopathy by correcting GH hypersecretion have been the subject of many studies. However earlier studies with traditional treatments are difficult to interpret and somewhat controversial because of the partial efficacy of these treatments and of the use, in the past, of less stringent criteria of cure.

Surgical treatment.

Surgical treatment of acromegaly was able to produce a significant decrease in plasma volume (32, 33) from 49.2±2 to 35.8±2 mL/kg body weight. Surgical treatment was suggested to improve cardiac hypertrophy in a study by Mather *et al.*In a cross-sectional study, left ventricular mass was found to be normal or moderatly increased in patients considered as cured when compared with that of patients with persistently increased GH levels which remained higher (34). In another study, acromegalic patients showed very heterogenous responses to surgical treatment. There was no clear difference between patients considered as cured and patients who were not cured. In fact, limited improvement of cardiac function was seen in 30 % of the cured patients. By contrast, cardiac function of patients who were not cured deteriorated and 3 out of 4 died (35). Surgical treatment of acromegaly was also found to produce a 30-40 % decrease in cardiac index, as measured by right heart catheterization (36).

Anecdotal case reports of acromegalic patients with cardiac congestive failure treated surgically have been published. After successful transsphenoidal hypophysectomy followed by biochemical remission, cardiac output and functional capacity improved in a 54-year old acromegalic patient (10) and in a 52-year old acromegalic patient (37) but the follow-up was short in these two studies (2 months to 2 years). More prolonged follow-up may be associated with recurence of GH hypersecretion leading to recurrence of cardiac failure and to an unfavorable outcome (37).

Radiation therapy.
The effects of radiation therapy on cardiovascular consequences of acromegaly have been disappointing. Despite a 60% decrease in GH levels during the 3-17-year follow-up of their patients, Baldwin *et al.* found that the cardiovascular events (myocardial infarction, dysrhythmias, hypertension, heart failure….) increased significantly in prevalence during this period (38). Such an unfavourable outcome despite reduction in GH levels by radiotherapy has also been reported by McGuffin *et al.* (19). The lack of improvement in heart disease in these studies probably reflects the low rate of biochemical remission achieved with radiotherapy and supports evidence for persisting or worsening of cardiovascular disease and/or cardiac risk factors until biochemical remission occurs.

Bromocriptine.
In a case report published in 1980, a decrease in GH levels from 74 to 7 µg/L obtained with bromocriptine in a 64-year old acromegalic woman with subclinical heart disease, was reported to be able to produce a decrease in interventricular septum thickness from 17 to 11mm and an improvement of ejection fraction. However this was a short-term study and no information concerning long-term follow-up is available in this patient (39).

Conclusion.
Inconsistencies in cardiovascular complications course after traditional treatment of acromegaly such as surgery, radiation therapy or bromocriptine may be due to multiple reasons : partial success of treatment, short observation period and inclusion of patients with end-stage disease (9, 40).

CARDIOVASCULAR EFFECTS OF SOMATOSTATIN ANALOGUES IN ACROMEGALY.

The discovery of the somatostatin analogs has provided a new approach to acromegaly treatment. In reducing GH and IGF-1 levels in 80% of cases and in normalizing them in 40-60% of cases, these drugs allow precisely the effects of GH hypersecretion correction on heart and vessels to be reassessed.

Effects of somatostatin analogs on cardiovascular parameters in patients with early or intermediate stages of acromegalic cardiac disease.
Plasma Volume. After somatostatin analog treatment, decrease in GH levels is associated with a fall in mean plasma volume (33, 41). However we were unable to show, at the individual level, that " normalization " of GH (< 2.5 µg/l) and IGF-1 levels was strictly associated with a normalization of volemia. Indeed, only 60 % of patients who achieved

normal GH and IGF1 levels had normalized plasma volume (33, 41, 42, our unpublished results).

Cardiac Index. Whether assessed by right heart catheterization (41) of by echography (43, 44), the cardiac index is a good indicator of the hyperkinetic heart syndrome. Mean cardiac index was shown to decrease during treatment with octreotide (41, 43, 44). According to the study by Lim *et al.*, this decrease was limited to acromegalic patients with left ventricular hypertrophy (LVH) on echocardiography (from 4.7±0.7 to 3.8±0.6 L/min • m^2, $p<0.05$, before treatment and 2 months after the start of octreotide), while it was not modified (3.3±0.3 to 2.8±0.2 L/min • m^2, p=ns) in patients without LVH (43). Thus, increased cardiac index is not universal in acromegalic patients and cannot be used as a sensitive marker for assessing cure of acromegaly. Moreover, when reviewing individual data in our patients, we found that no patient of the " improved " group (those whose reduction in GH levels exceeded 50 % from basal levels, without achieving normalization) and 41 % of the " well-controlled" group (GH levels less than 2 µg/L and normal IGF-1 levels) had normalized cardiac index 3 months after the beginning of octreotide treatment (41, 42, 45, our unpublished results).

Hypertension. In most of the studies, average blood pressure is not modified by somatostatin analogue treatment. This is probably due to the fact that, in the majority of the recent studies, acromegalic patients were selected on the basis of their absence of overt hypertension, in order to assess specific effects of analogues on cardiac structure and function, independently of comorbidity such as hypertension or diabetes mellitus... (18, 43, 45-48). At the individual level, a rate of persisting hypertension as high as 20–30 % in patients considered as cured (11, 42, 49) was reported. This persisting hypertension is not related to the level of GH achieved. It may be the consequence of vascular irreversible damage when the disease was active, or, alternatively, to the presence of a latent, independent, " essential " hypertension, which may be present in a subset of patients. Thus blood pressure levels cannot be used as a sensitive marker of good control of acromegaly after treatment.

Myocardial hypertrophy. Myocardial hypertrophy improves after effective treatment of acromegaly as demonstrated by the fall in mean left ventricular mass, interventricular septum thickness and left ventricular posterior wall thickness (11, 18, 43, 46-51) (Table 2). This effect is very rapid as it may be observed only one week after the beginning of treatment (43). Obviously, the effect is greater in patients having LVH as compared with patients having not LVH before treatment (43) (Figure 1).

However individual responses of acromegalic patients to somatostatin analogs are very variable. On one hand, reversal of cardiac hypertrophy was not obtained in all cured patients, since 25-30 % (versus 70-75 % before treatment) kept left ventricular hypertrophy (LVH) (18, 47) ; one may underline that the age of the patient is probably of concern for a reversal of LVH, as shown by Baldelli *et al.* who demonstrated that all their LVH patients less than 45-year old reverse LVH as compared with only 20 % of the older ones (18). On the other hand, even in patients unable to normalize GH and IGF-1 levels during somatostatin analogs, a decrease in LV mass may be observed (whether or not patients had LVH before treatment) (43).

In conclusion, myocardial hypertrophy generally improves during somatostatin analogs treatment. However, because LVH may be absent in some acromegalic patients, because total reversal may not be achieved even after biochemical cure, because LVH may be related to other independent factors such as hypertension, one cannot use reliably this morphological marker in the assesment of cure for acromegaly in an individual.

Table 2. Effects of long term treatment with somatostatin analogues (octreotide sc or lanreotide LP im) on morphological and systolic function parameters in acromegalic patients, according to various studies reported in the literature. LVH, left ventricular hypertrophy; B, before; D, during treatment with somatostatin analogues; LVM, left ventricular mass; IVST, interventricular septum thickness; LVPWT, left ventricular posterior wall thickness; FS, fractional shortening; LVEF, left ventricular ejection fraction. *, significant change between (D) and (B); ns, non significant change.

Author (year)		Lim (1992)		Merola (1993)	Tokgo-zoglu (1994)	Baldelli (1999)	Hradec (1999)
(ref)		(43)		(47)	(46)	(18)	(48)
N		10 LVH+	6 LVH-	11	6	13	13
Duration		2m	2m	6m	6m	12m	18m
Treatment		Octreotide SC				Lanreotide LP IM	
GH	B	17±4	27±7	34±6.5	15.8±9	10.1±2.2	86±110
(µg/l)	D	2.3±0.7*	5±1*	4.6±0.9*	4±4 *	3.9±0.9 *	33±55 *
LVM	B	339±28	193±24	256±9	304±80	263±15	157±37
(g)	D	275±25*	195±15 ns	210±19*	175±47*	211.8±10 *	124±31 *
IVST	B	11.3±0.3	9.2±0.2	13.7±0.6	-	10.2±0.4	12.5±1
(mm)	D	10.3±0.3*	8.7±0.2 ns	12.2±0.6*	-	9.1±0.5 *	11.3±1 *
LVPWT	B	11 ±0.3	9.5±0.3	11.9±0.6	-	10 ±4.4	12.4±1
(mm)	D	10 ±0.0.4*	9.0±0.3 ns	10.9±0.6*	-	8.8±0.2 *	11.1±1 *
FS	B	42.2±2.6	37.2±2.4	38±1.5	-	-	-
(%)	D	37.3±3.9 ns	33.5±3.2 ns	37±1.2 ns	-	-	-
LVEF	B	-	-	68+1.7	68±9	68±2	60±10
(%)	D	-	-	68±1.6 ns	65±10 ns	71.9±2.1 ns	62±10 ns

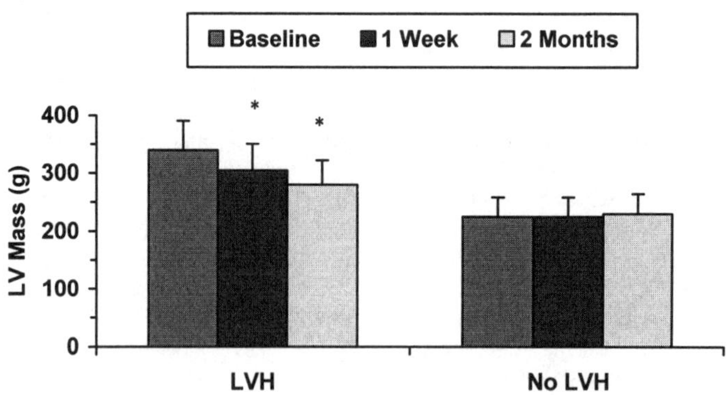

Figure 1. Effects of octreotide treatment on left ventricular mass (LV Mass))in acromegalic patients with and without LVH. * P < 0.01 compared with baseline. Adapted from Lim et al. (43).

Systolic function. As detailed in Table 2, somatostatin analog treatment did not modify significantly fractional shortening or left ventricular ejection fraction (LVEF) (46-48) assessed by echocardiography. Thus, decrease in left ventricular mass did not impair systolic function at rest.

Echocardiographic studies performed in short-term octreotide iv infusion (24 h) have shown that a very rapid and significant increase in several systolic functional indexes at rest (shortening fraction, ejection fraction, circumferential shortening rate and stroke volume) occurred during octreotide treatment, without any modification in echocardiographic anatomic parameters in these patients with LVH (44).

However, impairment in systolic function may not be seen at rest, becoming evident only during physical exercise (17). Effects of long-term treatment with somatostatin analogs on systolic function are much more evident when studied during physical exercise using gated blood pool cardiac scintigraphy (52). One-year treatment with octreotide significantly increased LVEF both at rest (from 56.5±1.8 % to 66.5±2.2 %, $p < 0.01$) and after exercise (from 52.6±2.4 % to 67.1±1.7 %, $p < 0.01$). By contrast, in patients who had persistently high circulating GH and IGF-1 levels, LVEF was unchanged at rest but was significantly impaired after exercice. This was particularly true for young patients (less than 40-year of age). At an individual level it is interesting to note that only half of the patients recover normal LVEF response to exercise after long-term treatment with octreotide.

Diastolic filling parameters. They have been evaluated by echocardiography (11, 18, 44, 47, 48, 50) or by gated blood pool cardiac scintigraphy (52). In four out of seven studies, an improvement in diastolic filling was observed during treatment with somatostatin analogues, either octreotide (44, 47, 50) or lanreotide (18). In three studies, diastolic filling did not modify significantly (11, 48, 52).

The effect on diastolic dysfunction may be very rapid as it was seen as early as the 24th hour following the beginning of octreotide iv infusion in 10 patients, despite GH and IGF-1 remaining elevated (44). Abnormally high A velocity and A/E velocity ratios were significantly decreased from 0.73±0.04 and 1.3±0.04 to 0.63±0.03 m/s and 1.05±0.04, respectively, while E velocity, which was normal basally, remained unchanged (44). It must be pointed out that this diastolic dysfunction persisted, albeit improved, in the long-term (18, 47, 50). The persistent diastolic dysfunction despite normalization of GH/IGF-1 levels, provides evidence for unreliability of this marker for assessing cure of acromegaly.

Cardiopulmonary performance during exercise. Few studies have analyzed the effect of somatostatin analogues on exercise performance. Exercise time on the treadmill which was 8.5±2.8 min at baseline improved significantly to 11.7±2.5 min after six months treatment with octreotide in 6 acromegalic patients (46). Similar results were reported by Colao *et al.* after one year treatment of octreotide, in the group of patients whose GH/IGF-1 levels had been normalized while exercise time remained limited to 7.6±0.4 minutes in patients whose GH/IGF-I levels were not normalized by octreotide (17). Exercice workload increased significantly from 80.8±6.4 to 92.3±5.9 W in the group of patients with normal GH levels while it remained stable (89.9±5.5 to 84.4±4.6 W) in the patients with persistently high GH/IGF-1 levels during octreotide treatment (17).

The effect of octreotide on cardiopulmonary performance seems to be very rapid, as demonstrated by Giustina *et al.* who reported that after 24-hour octreotide iv infusion, at anaerobic threshold, workload (69.6±8.9 W vs 54.6±7.3 W, basally), VO_2 (1179±127 mL/min vs 1036±114 mL/min, basally), minute-ventilation (30.4±3.5 L/min, vs 27.8±3 L/min, basally), and VCO_2 (1031±132 mL/min, vs 868±107 mL/mn, basally) increased. The same was true at maximal exercice. Once again it must be pointed out that if exercise capacity at anaerobic threshold after octreotide in acromegalic patients was not significantly different from that in normal subjects, this was not the case for maximal exercice which remained lower than in normal subjects (44).

Peripheral vascular parameters. We recently found decreased regional blood flow, at the level of the brachial artery, in active acromegalic patients when compared to normal controls, suggesting an heterogenous distribution of cardiac output in acromegaly (45). However, 3 months after treatment of acromegaly, we were unable to demonstrate any significant change in regional blood flow, probably due to the important interindividual variability and to the heterogenous response to treatment in this group of patients (our unpublished results).

This heterogenous hemodynamics in acromegaly prompted us to study endothelial function in these patients. By measuring local blood flow at the levels of the palm and the dorsum of the hand by laser-doppler we found that non specific vasodilatation (heating) was similar in acromegalic patients and in controls, reflecting normal vascular smooth cell relaxation in acromegaly. Impaired endothelium-dependent vasodilatation (mediated by no and studied by postocclusive hyperhemia) and increased sympathetic vasoconstriction (assessed by cold stress) were observed in acromegaly (53). Preliminary results indicate a trend for reversal of these abnormalities after effective treatment of acromegaly. However, individual responses as a function of hormonal responses have not yet been precisely analyzed (our unpublished results).

Conclusion. As regards cardiovascular parameters, correction of GH hypersecretion is associated with an improvement in all, suggesting that the decrease in cardiovascular morbidity and mortality may be very dependent on hormonal levels. In general, normalization of GH levels in acromegalic patients adds further benefits to the simple decrease (without normalization of GH/IGF-1 levels). However, even in the presence of prolonged normal GH/IGF-1 levels, many cardiovascular abnormalities remain, albeit at a lesser degree, addressing the question of total reversibility of acromegalic cardiomyopathy. The persistent cardiovascular abnormalities may be related to the importance of interstitial fibrosis described on histologic specimens (see above).

Effects of somatostatin analogs on cardiac function in patients with end-stage acromegalic cardiac disease.

In patients with acromegalic end-stage cardiac disease, treatment with somatostatin analogs has proved to rapidly improve cardiac function (26, 41). In 3 patients with congestive cardiac failure due to acromegalic cardiomyopathy, an improvement (24-51 %) in stroke volume and in cardiac index was observed, leading to normalization in filling pressures during octreotide treatment (Figures 2 and 3). This improvement was sustained up to 3 years (41).

These results were confirmed by Legrand *et al.* in a study of a 59-year old acromegalic patient with a severe dilated cardiomyopathy in whom myocardial biopsy before octreotide revealed signs of myocardial cell degeneration (extensive myofibrillolysis and areas of myocytolysis) with enlargement of myocardial fibers and presence of interstitial fibrosis (26). Improvement in cardiac function with octreotide allowed to perform surgical removal of a pituitary microadenoma which completely cured the hypersecretion of GH. Myocardial biopsy performed 4 months after surgery displayed that myocardial cell degeneration has became scarce and confined to small areas.

By contrast, Leclercq *et al.* in 1991 reported the case of an acromegalic patient with severe cardiac congestive failure due to acromegaly who worsened cardiac function during octreotide therapy, leading finally to propose heart transplantation (25, 27).
These apparently discordant results may be due to the variable severity of myocardial damage in these patients. Irreversible myocytolysis and extensive fibrosis which occur at an advanced stage of the disease makes recovery unlikely. At this stage octreotide may be detrimental because of its potential negative inotropic effect.

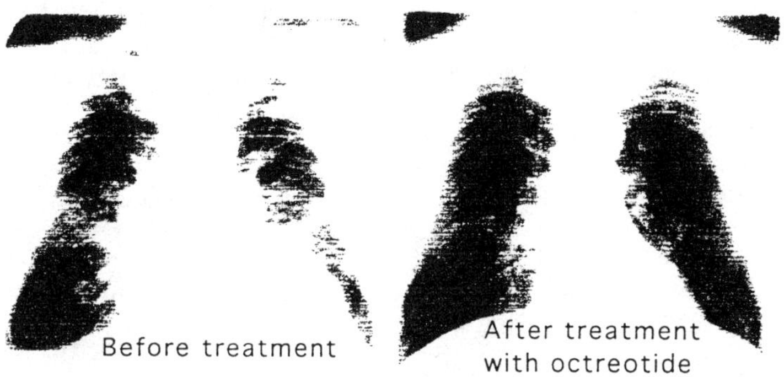

Figure 2. Effects of octrreotide treatment in an acromegalic patient with congestive heart failure. Note the reduction in cardio-thoracic index after 3 months of treatment.

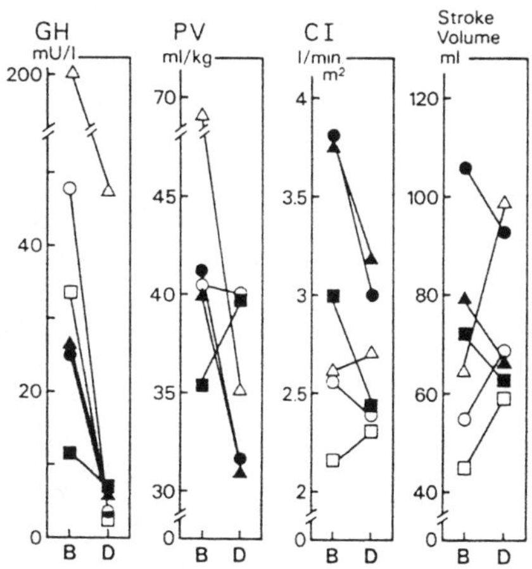

Figure 3. Changes in GH plasma levels, plasma volume (PV), cardiac index (CI) and stroke volume in 4 acromegalic patients without cardiac failure (closed symbols) and in 3 acromegalic patients with cardiac failure (open symbols). B=before treatment; D=during treatment. Reproduced with permission of the publisher from Chanson et al. (41).

GENERAL CONCLUSIONS

The effective treatment of acromegaly is associated with a reduction in left ventricular hypertrophy, an improvement in diastolic and systolic function and, a likely decrease in the excess cardiovascular mortality risk usually observed in untreated or insufficiently treated acromegaly. However, effective treatment is not always capable to totally reverse cardiac abnormalities. This raises the question of the determinants of this irreversibility. Young age of the patient, short duration of the disease, and absence of combined risk factors for cardiovascular disease, such as hypertension or diabetes or, maybe, sleep apnea syndrome, currently represent the best chances for the cardiomyopathy to reverse at a significant degree and sometimes for the heart to return to normal morphology and function. The role of interstitial fibrosis and apoptosis in the ability for this process to reverse is likely but needs futher to be elucidated.
Maximal lowering of GH/IGF-I levels has proved to improve cardiac prognosis in acromegaly. However, the issue of the safe GH levels which have to be achieved with an aggressive treatment of acromegaly needs also to be addressed. Indeed, epidemiological evidence showing increased mortality rate in hypopituitary patients with GH deficiency have accumulated in the few past years (54-56). Elsewhere, epidemiological studies conducted in the normal population, have also shown that lower as well as higher quartiles of normal GH levels were associated with increased cardiovascular risk (57).

Acknowledgements.
I wish to thank Pr José Timsit for his active and enthusiastic collaboration during many years. I also want to thank Drs Denis Chemla and Pierre Démolis, for their expert and continuous support in the area of vascular investigation.

REFERENCES

1. Wright AD, Hill DM, Lowy C, Fraser TR. Mortality in acromegaly. Q J Med1970;39:1-16.
2. Alexander L, Appleton D, Hall R, Ross WM, Wilkinson R. Epidemiology of acromegaly in the Newcastle region. Clin Endocrinol 1980;12:71-79.
3. Nabarro JD. Acromegaly. Clin Endocrinol 1987;26:481-512.
4. Bengtsson B-A, Eden S, Ernest I, Oden A, Sjögren B. Epidemiology and long-term survival in acromegaly. Acta Med Scand 1988;223:327-335.
5. Bates AS, Van't Hoff W, Jones JM, Clayton RN. An audit of outcome of treatment in acromegaly. Q J Med 1993;86:293-299.
6. Rajasoorya C, Holdaway IM, Wrightson P, Scott DJ, Ibbertson HK. Determinants of clinical outcome and survival in acromegaly. Clin Endocrinol 1994;41:95-102.
7. Shimatsu A, Yokogoshi Y, Saito S, Shimizu N, Irie M. Long-term survival and cardiovascular complications in paatients with acromegaly and pituitary gigantism. J Endocrinol Invest 1998;21:55-57.
8. Orme SM, McNally RJ, Cartwright RA, Belchetz PE. Mortality and cancer incidence in acromegaly: a retrospective cohort study. United Kingdom Acromegaly Study Group. J Clin Endocrinol Metab 1998;83:2730-2734.

9. Sacca L, Cittadini A, Fazio S. Growth hormone and the heart. Endocr Rev 1994;15:555-573.
10. Pepine CJ, Aloia J. Heart muscle disease in acromegaly. Am J Med 1970;48:530-534.
11. Lopez-Velasco R, Escobar-Morreale HF, Vega B, et al. Cardiac involvement in acromegaly: specific myocardiopathy or consequence of systemic hypertension? J Clin Endocrinol Metab 1997;82:1047-1053.
12. Minniti G, Jaffrain-Rea ML, Moroni C, et al. Echocardiographic evidence for a direct effect of GH/IGF-I hypersecretion on cardiac mass and function in young acromegalics. Clin Endocrinol 1998;49:101-106.
13. Melmed S. Tight control of growth hormone: an attainable outcome for acromegaly treatment [editorial]. J Clin Endocrinol Metab 1998;83:3409-3410.
14. Swearingen B, Barker FGn, Katznelson L, et al. Long-term mortality after transsphenoidal surgery and adjunctive therapy for acromegaly. J Clin Endocrinol Metab 1998;83:3419-3426.
15. Abosch A, Tyrrell JB, Lamborn KR, Hannegan LT, Applebury CB, Wilson CB. Transsphenoidal microsurgery for growth hormone-secreting pituitary adenomas: initial outcome and long-term results. J Clin Endocrinol Metab 1998;83:3411-3418.
16. Melmed S, Jackson I, Kleinberg D, Klibanski A. Current treatment guidelines for acromegaly. J Clin Endocrinol Metab 1998;83:2646-2652.
17. Colao A, Cuocolo A, Marzullo P, et al. Impact of patient's age and disease duration on cardiac performance in acromegaly: a radionuclide angiography study. J Clin Endocrinol Metab 1999;84:1518-1523.
18. Baldelli R, Ferretti E, Jaffrain-Rea ML, et al. Cardiac effects of slow-release lanreotide, a slow-release somatostatin analog, in acromegalic patients. J Clin Endocrinol Metab 1999;84:527-532.
19. McGuffin WL, Jr., Sherman BM, Roth F, et al. Acromegaly and cardiovascular disorders. A prospective study. Ann Intern Med 1974;81:11-18.
20. Rossi L, Thiene G, Caregaro L, Giordano R, Lauro S. Dysrhythmias and sudden death in acromegalic heart disease. Chest 1977;72:495-498.
21. O'Keefe JC, Wiseman JC, Wilmshurst EG. Dilated cardiomyopathy in acromegaly. Aust NZ J Med 1985;15:79-80.
22. Hirsch EZ, Sloman JG, Martin FI. Cardiac function in acromegaly. Am J Med Sci 1969;257:1-8.
23. Hradec J, Marek J, Kral J, Janota T, Poloniecki J, Malik M. Long-term echocardiographic follow-up of acromegalic heart disease. Am J Cardiol 1993;72:205-210.
24. Lie JT. Pathology of the heart in acromegaly: anatomic findings in 27 autopsied patients. Am Heart J 1980;100:41-52.
25. Leclerq F, Fille A, Albat B, Bringer J, Grolleau R, Jaffiol C. Congestive heart failure worsening with octreotide in acromegalic patient [letter]. Lancet 1991;338:1272-1273.
26. Legrand V, Beckers A, Pham VT, Demoulin JC, Stevenaert A. Dramatic improvement of severe dilated cardiomyopathy in an acromegalic patient after treatment with octreotide and trans-sphenoidal surgery. Eur Heart J 1994;15:1286-1289.
27. Albat B, Leclercq F, Serre I, et al. Heart transplantation for terminal congestive heart failure in an acromegalic patient. Eur Heart J 1993;14:1572-1575.
28. Frustaci A, Chimenti C, Setoguchi M, et al. Cell death in acromegalic cardiomyopathy. Circulation 1999;99:1426-1434.

29.Prysor-Jones RA, Jenkins JS. Effect of excessive secretion of growth hormone on tissues of the rat, with particular reference to the heart and skeletal muscle. J Endocrinol 1980;85:75-82.

30.Timsit J, Riou B, Bertherat J, et al. Effects of chronic growth hormone hypersecretion on intrinsic contractility, energetics, isomyosin pattern, and myosin adenosine triphosphatase activity of rat left ventricle. J Clin Invest 1990;86:507-515.

31.Peterson ME, Taylor RS, Greco DS, et al. Acromegaly in 14 cats. J Vet Int Med 1990;4:192-201.

32.Deray G, Rieu M, Devynck MA, et al. Evidence of an endogenous digitalis-like factor in the plasma of patients with acromegaly. N Engl J Med 1987;316:575-580.

33.Deray G, Chanson P, Maistre G, et al. Atrial natriuretic factor in patients with acromegaly. Eur J Clin Pharm 1990;38:409-413.

34.Mather HM, Boyd MJ, Jenkins JS. Heart size and function in acromegaly. Br Heart J 1979;41:697-701.

35.Hayward RP, Emanuel RW, Nabarro JD. Acromegalic heart disease: influence of treatment of the acromegaly on the heart. Q J Med 1987;62:41-58.

36.Passa P, Masquet C, Cophignon J, Gourgon R, Bouvrain Y. The heart in acromegaly. Haemodynamic study. Arch Mal Coeur Vaiss 1973;66:1517-1523.

37.Potier JC, Simon P, Vautier P, Foucault J-P. Cardiac output in evolutive acromegaly with cardiac failure. Nouv Presse Med 1977;6:2162.

38.Baldwin A, Cundy T, Butler J, Timmis AD. Progression of cardiovascular disease in acromegalic patients treated by external pituitary irradiation. Acta Endocrinol (Copenh) 1985;108:26-30.

39.Luboshitzki R, Barzilai D. Bromocriptine for an acromegalic patient. Improvement in cardiac function and carpal tunnel syndrome. JAMA 1980;244:1825-1827.

40.Pinzone JJ, Katznelson L, Klibanski A. Acromegalic heart disease. The Endocrinologist 1998;8:273-278.

41.Chanson P, Timsit J, Masquet C, et al. Cardiovascular effects of the somatostatin analog octreotide in acromegaly. Ann Intern Med 1990;113:921-925.

42.Houdoin L. Complications cardiovasculaires de l'acromégalie; Thesis 1993, Paris.

43.Lim MJ, Barkan AL, Buda AJ. Rapid reduction of left ventricular hypertrophy in acromegaly after suppression of growth hormone hypersecretion. Ann Intern Med 1992;117:719-726.

44.Giustina A, Boni E, Romanelli G, Grassi V, Giustina G. Cardiopulmonary performance during exercise in acromegaly, and the effects of acute suppression of growth hormone hypersecretion with octreotide. Am J Cardiol 1995;75:1042-1047.

45.Chanson P, Meignien J-L, Del Pino M, et al. Decreased regional blood flow in patients with acromegaly. Clin Endocrinol 1998;49:725-731.

46.Tokgozoglu SL, Erbas T, Aytemir K, Akalin S, Kes S, Oram E. Effects of octreotide on left ventricular mass in acromegaly. Am J Cardiol 1994;74:1072-1074.

47.Merola B, Cittadini A, Colao A, et al. Chronic treatment with the somatostatin analog octreotide improves cardiac abnormalities in acromegaly. J Clin Endocrinol Metab 1993;77:790-793.

48.Hradec J, Kral J, Janota T, et al. Regression of acromegalic left ventricular hypertrophy after lanreotide (a slow-release somatostatin analog). Am J Cardiol 1999;83:1506-1509, A1508.

49.Colao A, Ferone D, Cappabianca P, et al. Effect of octreotide pretreatment on surgical outcome in acromegaly. J Clin Endocrinol Metab 1997;82:3308-3314.

50.Pereira JL, Rodriguez-Puras MJ, Leal-Cerro A, et al. Acromegalic cardiopathy improves after treatment with increasing doses of octreotide. J Endocrinol Invest 1991;14:17-23.
51.Thuesen L, Christensen SE, Weeke J, Orskov H, Henningsen P. The cardiovascular effects of octreotide treatment in acromegaly: an echocardiographic study. Clin Endocrinol 1989;30:619-625.
52.Colao A, Cuocolo A, Marzullo P, et al. Effects of 1-year treatment with octreotide on cardiac performance in patients with acromegaly. J Clin Endocrinol Metab 1999;84:17-23.
53.Chanson P, Maison P, Demolis P, Young J, Giudicelli JF, Schaison G. Endothelial dysfunction and increased sympathetic vasoconstriction in normotensive acromegalic patients. The Endocrine Society's 81st Annual Meeting, June 12-15, 1999. San Diego, Cal; (Abst P3-649).
54.Rosen T, Eden S, Larson G, Wilhelmsen L, Bengtsson BA. Cardiovascular risk factors in adult patients with growth hormone deficiency. Acta Endocrinol 1993;129:195-200.
55.Bülow B, Hagmar L, Mikoczy Z, Notdström C-H, Erfurth EM. Increased cerebrovascular mortality in patients with hypopituitarism. Clin Endocrinol 1997;46:75-81.
56.Bates AS, Van't Hoff W, Jones PJ, Clayton RN. The effects of hypopituitarism on life expectancy. J Clin Endocrinol Metab 1996;81:1169-1172.
57.Maison P, Balkau B, Simon D, et al. Growth hormone as a risk for premature mortality in healthy subjects: data from the Paris prospective study. Br Med J 1998;316:1132-1133.

7

GROWTH HORMONE SECRETION IN CONGESTIVE HEART FAILURE

Filippo Manelli[1], Maurizio Volterrani[2], Roberto Lorusso[3], Giuseppe Romanelli[1], Andrea Giustina[1]

[1]Endocrine and Geriatric Sections, Dept of Internal Medicine, University of Brescia; [2]Fondazione "S. Maugeri", Cardiology Dept., Gussago Medical Centre, Brescia; [3]II Cardiac Surgery, Spedali Civili di Brescia; Italy

1. PHYSIOLOGICAL REGULATION OF GROWTH HORMONE SECRETION

Growth hormone (GH) secretion is under the control of two hypothalamic peptides (1). GH-releasing hormone (GHRH), a peptide of 44 amino acids, which stimulates GH release and somatostatin, which exists both as 14 and 28 amino acid peptides, which inhibits GH secretion. GH secretion is regulated by negative feed-back and neural control mechanisms. Both GH and IGF-1 inhibit GH secretion after intraventricular injection by promoting hypothalamic somatostain release. Presumably, physiological concentration of GH and IGF-1 reaching the hypothalamus in the bloodstream act in the same way (increase in somatostatin tone). In addition, IGF-1 may act directly on the pituitary to inhibit GHRH-stimulated secretion of GH. GH secretion can be augmented or inhibited by a number of neurogenic, metabolic, and hormonal influences (Figure 1). In the adult, the diurnal pattern of GH secretion has been characterized by obtaining blood samples every 20 or 30 minutes throughout a 24-hour period under non stressful conditions. During most of the day, plasma GH levels of normal adults are less than 5 ng/mL, with one or two sharp spikes three to four hours after meals. The most consistent period of GH secretion for both children and young adults occurs about one hour after the onset of deep sleep (2).

2. GH SECRETION IN CONGESTIVE HEART FAILURE

Congestive heart failure (CHF) is a clinical syndrome with an overall poor prognosis, despite the advances in drug treatment. The only effective treatment for end-stage heart failure remains heart transplantation, which is limited by a shortage of donor organs; therefore, new therapeutic strategies are under investigation. CHF is characterized by left ventricular impairment which leads to secondary changes in other organs with consequent symptoms like dyspnea, muscular fatigue and exercise intolerance. While in the past heart failure has been considered mainly an hemodynamic disorder, in the last decade the attention has focused on the muscular and hormonal changes in this condition.

The role of the GH/IGF-1 axis in heart failure has been carefully investigated only in the last few years. In fact, spontaneous GH hypersecretion or hyposecretion have been associated to different stages of heart dysfunction in humans (3). Overall, these data suggested that the

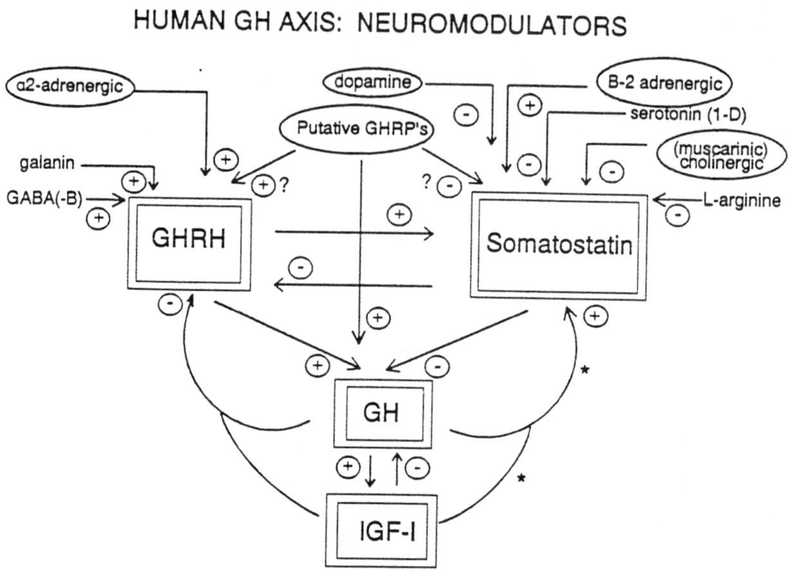

Figure 1. Summary representation of the putative roles of the principal neuropeptides and neurotransmitters that supervise GH secretion via GHRH or somatostatin (SS) or by acting on the pituitary gland (GH) in human. Asterisks denote that two more loci of action are recognized. Not shown are numerous other metabolic and hormonal effectors that also act via multiple pathways, e.g., IGF-1, sex steroids, age, glucocorticoids, diabetes mellitus, obesity, T_4, etc. An unproved role for putative (as yet unidentified) GHRP-like endogenous ligand is also noted, given that receptors for GHRP ligands are expressed in the hypothalamus and pituitary gland.

pituitary could have a role in the endocrine compensatory response to heart failure. GH has the ability to exert its action directly at the tissue level interacting with its own receptors even if most of the biological effects of GH are known to be mediated by the hepatic and peripheral production of IGF-1 (4). GH secretion is pulsatile and is modified by external stimuli and endogenous neural rhythms, and by the feedback effects of several circulating hormones (5). Moreover, plasma half-life of GH is relatively short. Therefore, an adequate schedule of repetitive and prolonged blood sampling must be implemented to obtain significant and useful quantitative information concerning pulsatile GH release (6). Conversely, since IGF-1 has much longer half-life than GH and circulating IGF-1 levels do not significantly change through the day, a single sample assay may give a reliable measure of IGF-1 (4). Measurement of IGF-1, in turn, gives a reasonable idea of the 24-h GH secretory rate and is routinely used in the assessment of the activity of the GH axis in pathological conditions such as pituitary dwarfism and acromegaly (7). The limitations of the clinical use of the IGF-1 assay as marker of GH secretion are mainly two: first, IGF-1 synthesis is not only regulated by GH but also by nutrient supply and by other hormones (8); second, it has been reported that low IGF-1 levels in the presence of high-normal GH Synthesis may reflect a peripheral resistance to GH action (9).

Initially, we assessed the pattern of spontaneous nocturnal GH secretion in a group of patients with dilated cardiomyopathy and varying degrees of cardiac failure (10). All of the patients had normal baseline blood pressure levels (111/70±5/2 mmHg), were under conventional medical treatment for heart failure, were assigned either to New York Heart Association class III (10 patients) or IV (2), and had dilated left ventricle (end-diastolic volume > 200 ml) and ejection fraction < 30 % (mean 22.7±2.5%, range 15% to 29%). Nine volunteers with normal ejection fraction served as controls. Each subject spent one night in our sleep unit after undergoing echocardiographic evaluation (10). Lights were turned off at 10 PM and turned on at 6 AM. Subjects rested in bed at 10 PM, and a catheter was inserted into a forearm vein. The first blood sample for GH and cortisol was taken at 10:30 PM. Subsequent blood samples were taken at 30-minute intervals until 6 AM the next day. A positive significant ($p < 0.05$) correlation between mean nocturnal GH and ejection fraction value (r=0.813) in patients with dilated cardiomyopathy was found. Conversely, no significant correlation was found between ejection fraction and spontaneous GH secretion in normal subjects. In the patients with ejection fraction of ≤ 20%, all parameters of spontaneous GH secretion and the mean ejection fraction were significantly reduced compared to parameters in the subgroup of patients with dilated cardiomyopathy and ejection fraction of > 20% and to normal subjects. The subgroup of patient with ejection fraction > 20% showed a slight, even if not significant, increase in the various parameters of spontaneous GH secretion as compared to normal subjects. No significant differences in the various parameters of spontaneous nocturnal cortisol secretion were observed between groups. This was the first attempt to correctly assess spontaneous nighttime GH secretory status in patients with cardiac failure. The use of this new and correct method suggested that the severity of cardiac failure in patients with dilated cardiomyopathy is strictly linked to a decrease in circulating GH levels at night. It was hypothesized that GH may be involved in the regulatory responses to heart failure and that failure of this compensatory mechanism may cause a significant worsening of ventricular function.

In a second study, in twelve male patients with chronic CHF (ischemic or idiopathic), we evaluated (11) nocturnal GH secretion and IGF-1 levels at baseline and the GH and IGF-1 response to 24 hours human recombinant GH intravenous infusion.

Aim of this study was to assess both if there could be any correlation between GH and/or IGF-1 secretion and hemodynamic (therefore, more accurate than the echocardiographic ones evaluated in the previous study) parameters in CHF, and if spontaneous GH secretion rate or baseline IGF-1 may be predictors and/or determinants of the acute effects of continuous intravenous infusion of GH on hemodynamic parameters in patients with CHF. The study involved twelve male patients with chronic CHF (ischemic or idiopathic), no change in signs and symptoms in the last three months, end diastolic left ventricular dimension > 65 mm, left ventricular ejection fraction below 35% (mean 20.7±6.6%), as assessed by two-dimensional echocardiography; chronic oral therapy with digoxin, diuretics and vasodilators; a peak exercise oxygen consumption ≤ 16 mL/Kg/min (mean 10.2±3.1 mL/Kg/min), and sinus rhythm. Patients with diabetes mellitus, hypertension, chronic alcoholism, severe renal (serum creatinine > 2.5 mg/dL) and hepatic (serum albumin < 3.5 g/L; pseudocholinesterase < 3000 IU/L) failure were excluded from the study. Furthermore, we did not include in the study patients receiving either alpha and beta adrenergic blockers or other drugs acting on the central nervous system. All the patients entered the study after a

7-day clinical and therapeutical stabilization period in the hospital. The protocol of the hemodynamic study has been previously described (12). Briefly, the first 24 hours after right heart catheterization (from 8.00 a.m. of the first day of the study to 8.00 a.m. of the second day) were considered the control period; in fact, during the following 24 hours (from 8.00 a.m. of the second day of the study to 8.00 a.m. of the third day) all the patients underwent intravenous constant pump infusion (Abbott, Lifecare 5000, North Chicago, Illinois, USA) of 0.1 IU/Kg/24h recombinant human GH (Humatrope, Eli Lilly, Italy). Hemodynamic parameters were continuously monitored during the 48 hours of the study. The protocol of the endocrine part of the study was designed as follows. Blood samples for GH assay were taken every 20 min during the first night of the study (from 10 p.m. of the first day of the study to 6.00 a.m. of the second day of the study). Moreover, blood samples for GH assay were also taken during exogenous GH infusion (at 8.00, 10.00 and 12.00 a.m. of the second day of the study and then every four hours until 8.00 a.m. of the third day of the study). Blood samples for IGF-1 assays were taken at 8.00 a.m. of each of the three days of the study. Standard medical therapy for heart failure was continued unchanged throughout the study. Meals, which were standardized for each patient (8.6 MJ/day; 15% protein, 50% carbohydrates; 35% lipids) were served at 7.00 a.m., 12.00 a.m., 6.00 p.m., of each day of the study. No extra calories were allowed between meals. In our patients, the mean nocturnal GH level was 2.5±1.4 μg/L. Peak GH levels and GH AUC resulted to be 8.6±7.3 μg/L and 1467±1342.1 μg/L • 8h, respectively. All parameters of spontaneous nocturnal GH secretion and circulating IGF-1 levels showed a wide variability ranging from the low range of our normal population (patient # 4 and 5) to very pronounced spontaneous GH secretory activity (patient # 8 and 10). It was possible to notice that the 7 patients with higher mean spontaneous GH levels (>2 μg/L) tended to have lower circulating IGF-1 levels as compared to the 5 patients with apparently lower spontaneous GH secretory activity (mean nocturnal <2 μg/L): 143.8±49.5 vs 203.5±64.4 μg/L, p=0.09. Mean GH serum concentration increased, on the average, during infusion five fold as compared to the baseline level (from 2.5±1.4 to 13.6±8.3 μg/L) in our patients; IGF-1 rose of about 50% as compared to the pre-infusion level (from 168.7±61.6 to 248.3± 86.6 μg/L). A certain degree of variability in the effects of exogenous GH on serum GH levels and IGF-1 levels was also observed (13). Mean GH levels during the infusion ranged from 3.4 to 33.1 μg/L (from 5 to more than 10 fold increase). Moreover, IGF-1 levels after exogenous GH ranged from 129.9 to 403.9 μg/L (from 15 to 115% increase). Interestingly, the highest IGF-1 levels after the GH infusion were obtained in those patients with higher IGF-1 levels at baseline independently of the serum GH levels obtained during the infusion. In fact, the 6 patients with low baseline IGF-1 levels (<180 μg/L) showed a reduced increase in delta-IGF-1 levels (peak-baseline: 90.6±53.9) as compared to the 6 patients with baseline IGF-1 level >180 μg/L (68.6±29.2 μg/L). A negative correlation, however not reaching the statistical significance, has been found between mean nocturnal GH levels and baseline IGF-1 levels (r=-0.47, p=0.124). Interestingly, negative significant correlations have been found between mean nocturnal GH levels and both post infusion absolute (r=-0.67, p<0.05) and delta (post- pre-infusion) (r=-0.58; p<0.05) IGF-1 levels. No significant correlations have been found between several parameters of liver function (albumin, bilirubin and pseudo-cholinesterase) and mean nocturnal GH. However, baseline IGF-1 levels showed a negative significant correlation (r=-0.76, p<0.01) with total bilirubin and a positive correlation (r=0.72, p<0.01) with pseudo-cholinesterase (Figure 2c).

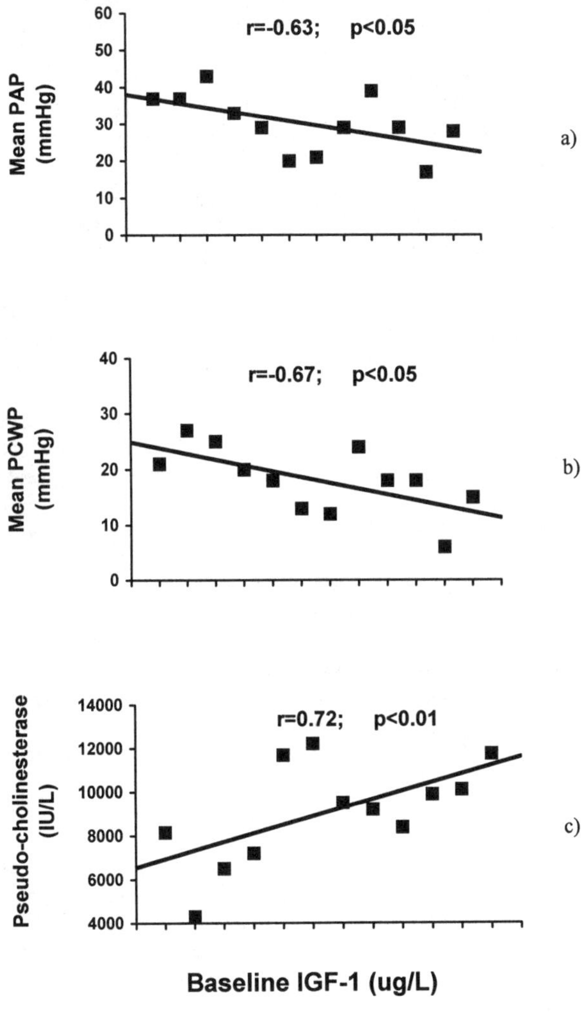

Figure 2. Correlation between baseline IGF-1, mean pulmonary arterial pressure (mPAP) (a), capillary wedge pressure (PCWP) (b) and pseudo-cholinesterase (c) after GH infusion.

No significant correlations between spontaneous or post infusion GH levels and the hemodynamic data, both at baseline and after GH infusion were observed. Baseline IGF-1 levels showed a significant, negative correlation with baseline mean PAP ($r=-0.68$, $p<0.05$)

and PCWP (r=-0.67, p<0.05) and a positive correlation with baseline cardiac index (r=0.71, p<0.05). Baseline IGF-1 levels showed also a significant negative correlation with absolute mean PAP (r=-0.63, p<0.05) (Figure 2a) and mean PCWP (r=-0.67, p<0.05) (Figure 2b) after GH infusion. Post GH infusion IGF-1 levels also negatively correlated with post GH infusion mean PAP (r=-0.50, p=0.09) and mean PCWP (r=-0.66, p<0.05). The positive correlation between either baseline or post-infusion IGF-1 and the post-infusion cardiac index (r=0.40 and 0.43, respectively) did not reach statistical significance.

These latter data give some interesting insights concerning the status of the GH axis in patient with CHF and an how it can affect the cardiovascular effects of exogenous GH. Interestingly, the cardiovascular effects of GH were marginally dependent on the baseline hemodynamic picture of the subjects (weak correlations between baseline and post-GH hemodynamic parameters). Conversely, the baseline endocrine picture of the subjects appeared to be of some importance in determining the cardiovascular responsivity to the GH infusion. In fact, patients with low circulating IGF-1 levels and spontaneously high GH levels (i.e. those characterized by a certain degree of "GH resistance") were likely to have less dramatic endocrine (increase in IGF-1) and hemodynamic responses to GH infusion. It seems unlikely that other factors may have contributed to the different IGF-1 response to GH infusion observed in our patients; in fact, nutritional and external conditions were maintained constant throughout the study. In the patients with low baseline IGF-1 a more compromised liver function, likely consequent to CHF, was found: in fact, baseline IGF-1 levels correlated negatively with serum total bilirubin and positively with serum pseudo-cholinesterase. It could be hypothesized either that impaired IGF-1 response to GH may be a biological marker of generalized tissue resistance to GH (and therefore that the cardiovascular tissue may be less responsive to the direct action of GH) or that the hepatic or cardiovascular synthesis of IGF-1 in response to GH could be reduced (and therefore the cardiovascular response to circulating or locally produced IGF-1 is impaired). In the clinical practice are already known pathological conditions characterized by genetic (Laron dwarfism) (9) or acquired (type 1 diabetes mellitus, renal insufficiency, starvation) (14-16) GH resistance: in all these conditions impaired IGF-1 synthesis has been demonstrated to decrease the biological effect of GH at the tissue levels.

If, in the clinical setting, low baseline circulating IGF-1 levels really represent a good predictor of a weak hemodynamic response to exogenous GH infusion, one should expect that we will be able to find with further studies the cut-off IGF-1 level under which GH administration has not significant acute hemodynamic effects. It can be hypothesized that a subset of patients with CHF and characterized by low baseline IGF-1 levels (and high GH) and blunted IGF-1 response to GH are characterized by peripheral GH resistance (11): peripheral GH resistance may impair cardiovascular response to exogenous GH according to an impaired hepatic (as suggested by the correlation between IGF-1 levels and liver function) or cardiac production of IGF-1 (it implicates a role for systemic or local IGF-1 in cardiovascular response to GH) (17) or an impaired GH action at cardiac level (it implicates a direct effects for GH on heart function).

Furthermore, these data suggest that all the studies on the cardiovascular effects of GH need to be designed carefully considering the baseline endocrine picture of the subjects and particularly the enrollement in these studies of patients with peripheral resistance to GH: in fact, our data suggest that those patients are unlikely to have all the expected beneficial hemodynamic effects from GH infusion.

3. CONCLUSIONS

Circulating GH and IGF-1 levels in patients with CHF suggest that, at least in some of them, peripheral GH resistance may occur. Therefore, on the basis of recent data giving some interesting insights concerning the possible determinants of the observed positive functional acute cardiovascular effect of GH, it should be carefully considered the baseline endocrine picture of the patients to select those who might have more clinical benefits from this new therapeutic option.

REFERENCES

1. Wehrenberg WB, Giustina A. Basic counterpoint: mechanisms and pathways of gonadal steroid modulation of growth hormone secretion. Endocr Rev 1992;13:299-308.
2. Giustina A, Veldhuis JD. Pathophysiology of the neuroregulation of growth hormone secretion in experimental animals and the human. Endocr Rev 1998;19:717-797.
3. Saccà L, Cittadini A, Fazio S. Growth hormone and the heart. Endocr Rev 1994;**15**:555-573.
4. Zapf J, Walter H, Froesch ER. Radioimmunological determination of insulin like growth factors I and II in normal subjects and in patients with growth disorders and extrapancreatic tumor hypoglycemia. J Clin Invest 1981;68:1321-1330.
5. Giustina A, Wehrenberg WB. The role of glucocorticoids in the regulation of growth hormone secretion. Trends Endocrinol Metab 1992;3:306-311.
6. Veldhuis JD. The neuroendocrine regulation and implications of pulsatile GH secretion: gender effects. The Endocrinologist 1995;5:198-213.
7. Giustina A, Zaltieri G, Negrini F, Wehrenberg WB. The pharmacological aspects of the treatment of acromegaly. Pharmacol Res 1996;34:1-22.
8. Giustina A, Wehrenberg WB. Growth hormone neuroregulation in diabetes mellitus. Trends Endocrinol Metab 1994;5:73-78.
9. Savage MO, Blum WF, Ranke MB, Postel-Vinay MC, Cotteril AM, Hall K. Clinical features and endocrine status in patients with growth hormone insensitivity (Laron syndrome). J Clin Endocrinol Metab 1993;77:1465-1471.
10. Giustina A, Lorusso R, Borghetti V, Bugari G, Misitano V, Alfieri O. Impaired spontaneous growth hormone secretion in severe dilated cardiomyopathy. Am Heart J 1996;131:629-632.
11. Giustina A, Volterrani M, Manelli F, et al. Endocrine predictors of the acute hemodynamic effects of growth hormone in congestive heart failure. Am Heart J 1999;137:1035-1043.
12. Volterrani M, Desenzani P, Lorusso R, D'Aloia A, Manelli F, Giustina A. Hemodynamic effects of intravenous growth hormone in congestive heart failure. Lancet 1997;349, 1067-1068.
13. Laursen T, Jorgensen JO, Jacobsen G, Hansen BL, Christiansen JS. Continuous infusion versus daily injections of growth hormone (GH) for 4 weeks in GH-deficient patients. J Clin Endocrinol Metab 1995;80:2410-2418.
14. Giustina A, Bossoni S, Cimino A, Pizzocolo G, Romanelli G, Wehrenberg WB. Impaired growth hormone (GH) response to pyridostigmine in type 1 diabetic patients with exaggerated GH-releasing hormone-stimulated GH secretion. J Clin Endocrinol Metab 1990;7:1486-1490.

15. Blum WF, Ranke MB, Kietzmann K, Tonshoff B, Mehls O. Growth hormone resistance and inhibition of somatomedin activity by excess of insulin-like growth factor binding protein in uremia. Ped Nephrol 1991;**5**:539-44.
16. De Marinis L, Folli G, D'Amico C, et al. Differential effects of feeding on the ultradian variation of the growth hormone (GH) response to GH-releasing hormone in normal subjects and patients with obesity and anorexia nervosa. J Clin Endocrinol Metab 1988;66:598-604.
17 Wetter U, Kupferschmid C, Lang D, Pentz S. Insulin-like growth factors and insulin increase the contractility of neonatal rat cardiocytes in vitro. Basic Res Cardiol 1988; 83:647-54.

8

THE GROWTH HORMONE AXIS IN CACHECTIC PATIENTS WITH CHRONIC HEART FAILURE: EVIDENCE FOR ACQUIRED GROWTH HORMONE RESISTANCE.

Stefan D. Anker[1,2], Claus-Dieter Pfaum[3], Mariantonietta Cicoira[2], Christian J. Strasburger[3]

[1]*Franz-Volhard-Klinik (Charité, Campus Berlin-Buch) at Max-Delbrück-Centrum for Molecular Medicine, Berlin, Germany;* [2]*Cardiac Medicine, NHLI, Imperial College School of Medicine, Dovehouse Street, London, UK;* [3]*Medizinische Klinik, Ludwigs Maximilian Universität München, München, Germany.*

INTRODUCTION

Chronic heart failure (CHF) is a clinical syndrome with an overall poor prognosis, despite the advances in drug treatment. CHF is characterized by left ventricular impairment which leads to secondary changes in other organs with consequent symptoms like dyspnoea, muscular fatigue and exercise intolerance. It has long been recognised that significant body wasting (i.e. weight loss) is also an important feature of severe CHF. This dates back to the time of Hippocrates (about 460 - 370 BC), when he wrote: "The flesh is consumed and becomes water, ... , the shoulders, clavicles, chest and thighs melt away. This illness is fatal, ..." (1). Like inflammation and fever; weight loss is frequently observed as one of the common patterns of response to sever disease. The presence of general weight loss in patients with CHF has been termed *cardiac cachexia*. The term cachexia is of Greek origin, derived from the words *kakos* (i.e. bad) and *hexis* (i.e. condition). Cachexia is one of the most visible and devastating consequences of human disease, similarly seen in a number of chronic diseases like cancer, acquired immunodeficiency syndrome (AIDS), thyrotoxicosis and rheumatoid arthritis. In malignant cancer, AIDS and chronic lung disease, cachexia is well known to be a sign of very poor prognosis. In principal it is thought to be related to loss of appetite (anorexia), anemia and metabolic abnormalities, that in turn are influenced by altered hormones and cytokines. Acquired GH resistance is a feature of severe catabolism and malnutrition in conditions of sepsis, surgery and critical illness (reviewed in (2,3). Biochemically it is defined as the presence of high GH but low IGF-I levels. GH resistance, i.e. GH insensitivity, is typical in Laron dwarfism, which results from a GH receptor mutation (4).

THE DEFINITION OF CARDIAC CACHEXIA

While in the past heart failure has been considered mainly a hemodynamic disorder, the study of metabolic, hormonal and immunological abnormalities has evolved during the last 10 years. The neurohormonal hypothesis postulates that although haemodynamic abnormalities initiate the symptoms of patients with CHF, they are not sufficient to explain the disease progression, and therefore other mechanisms may play a role in determining the worsening of the symptoms and ultimately the death of the patient (5). It has been shown that hormonal changes in CHF patients have no close relation to the conventional parameters of CHF severity, like left ventricular ejection fraction (LVEF), New York Heart Association (NYHA) functional class and exercise capacity, but rather with the development of cardiac cachexia (6). Very little is known about the transition from stable non-cachectic CHF to the development of cachexia. Even the definition of cardiac cachexia and the characteristics of the cachectic patient remain controversial. Cachexia is usually defined as a syndrome of severe weight loss. Therefore it has been suggested to use a weight loss as the main criterion to define weight loss. Cardiac cachexia can be diagnosed on the basis of documented non-intentional and non-edematous weight loss >7.5% of the pre-morbid normal weight occurring over a time period of >6 months (7). Using this definition, 16% of an unselected population of 171 CHF patient were found to be cachectic, and the presence of cachexia is a strong independent risk factor for mortality in patients with CHF, with a mortality in the cachectic cohort of 50% at 18 months as compared to 83% in non-cachectic CHF patients (8).

THE CAUSE OF CACHEXIA IN HEART FAILURE

When using documented weight loss as criterion to dichotomise CHF patients, it has been shown that cachectic CHF patients not only suffer from significant loss of lean tissue (ie, protein reserve) but also they show a grossly reduced fat tissue (ie, energy reserve) and bone mass (9,10), indicating an increased catabolism and increased resting metabolic rates that have been found as well (11). In 1964, Pittman and Cohen were the first to analyse the pathogenesis of cardiac cachexia (12). They thought the development of cellular hypoxia to be the leading pathogenic factor causing less efficient intermediary metabolism, therefore producing increased catabolism. Many have argued that cardiac cachexia is due to malnutrition and loss of appetite. Protein loosing enteropathy has been excluded in a study of cachectic CHF patients (13). Simple starvation and anorexia would mainly lead to a loss of fat tissue (which is not the case, see above) and they would also cause reduced plasma albumin levels. Yet, albumin as well as liver enzyme plasma levels are not decreased in the cachectic patients (6,14). Physical inactivity and deconditioning have been suggested to be important for the muscle atrophy observed in many patients with CHF (15), but histological evidence suggests that the atrophy in states of reduced activity is significantly different from the muscle atrophy observed in CHF (16). Therefore, anorexia, malabsorption or liver dysfunction, and physical inactivity are unlikely to be the major factor of cachexia development in CHF patients. We have proposed that the imbalance of catabolic and anabolic metabolic factors causes the development of cardiac cachexia (6).

FIRST STUDIES OF GH AND IGF-1 LEVELS IN CHF

Cross-sectional studies have suggested that neuroendocrine and immunological abnormalities are linked to the presence of body wasting. Neurohormones (17) and cytokines (18-20) can predict survival in CHF and are strongly related to the presence of cardiac cachexia. Comparing the features of cachectic and non-cachectic CHF patients to those of healthy subjects, it is mainly the cachectic patients who show raised plasma levels of catecholamines, cortisol, aldosterone and tumor necrosis factor-alpha (TNFα) and the highest plasma renin activity. Raised plasma levels of growth hormone (GH) have first been described by Anand et al. in 1989 (21). These patients had severe heart failure and were untreated at the time of assessment. When we investigated plasma levels of total GH in CHF patients (using the immuno radiometric assay from Nichols Institute Diagnostics, San Juan Capistrano, USA), we found them to be particularly high in cachectic CHF patients (mean 3.8 ng/ml, n=16), but low in non-cachectic patients (0.9±0.3 ng/ml, n=37) and healthy control subjects (1.1±0.4 ng/ml, n=16) (6). No other subgrouping (by NYHA class, LVEF, or peak VO_2) yielded significant differences between patient subgroups (Figure 1). Plasma levels of IGF-I were very similar between patient subgroups (Figure 1). As IGF-I is the main mediator of the anabolic effects of GH, the relationship between these two hormones has also been studied. The IGF-I to GH ratio was higher in non-cachectic CHF patients and controls compared to cachectics. As this ratio has a skewed distribution, the log-transformed ratios were compared statistically (controls: 2.89±0.25; non-cachectics: 3.00±0.16; cachectics: 2.03±0.22, p=0.014 vs controls and p=0.0014 vs non-cachectics) (6). As the increase in GH in our cachectic patients was not accompanied by an increase of IGF-I, this suggested the presence of GH resistance. Interestingly, when analysing the correlation between humoral parameters and the degree of weight loss in 16 cachectic patients, the 3 variables TNFα, log IGF-I /GH ratio and plasma testosterone together predicted 83.5% of the variation of the documented weight loss (in kg) and 84.7% of the variation of the relative weight loss (in %). Neither testosterone nor log IGF-I/GH significantly correlated with the body mass index or measures of weight loss in univariate analysis, but both parameters became (independently of each other) important after adjusting for the effect of TNFα.

THE GH-IGF1 AXIS IN CARDIAC CACHEXIA

Study protocol. Recently, we have extended these studies (22): a) to increase the number of patients in these investigations, and b) to assess some other important parameters of the GH - IGF-I axis in CHF patients. Lack of GH receptors could cause GH resistance. GH-binding protein (BP) is structurally identical to the GH receptor ectodomain and is thought to reflect the cellular GH receptor status (23,24). For the analysis of biologically intact GHBP a ligand immunofunctional assay has been developed (25). Additionally, we assessed plasma levels of the immunofunctionally intact form of GH, using an immunofunctional assay (26), and we studied levels of IGFBP3 using a non-isotopic sandwich immunoassay (27). A total of 98 subjects were investigated, 26 healthy controls (age 56±2 years) and 72 stable patients with CHF (61±1 years). The patients had a mean LVEF of 27±2% and a mean peak VO_2 of 16.5±0.7 ml/kg/min compared to 36.1±1.4 ml/kg/min in controls (p<0.0001). The CHF patients were on clinically optimal drug treatment, and no patient had severe renal failure (all serum creatinine levels <255 μmol/l). Twenty-one CHF patients were defined as cachectic (weight loss 12.5±1.5 kg [range 5.5 to 30 kg] over 2.8±0.6 years, i.e. 6.4±1.0 kg per year and 16.2±1.7 % of previous normal weight) The body mass index of these patients was 21.1±0.4 kg/m^2 compared to 26.6±0.5 kg/m^2 (p<0.0001) in 51 non-cachectic CHF patients. Comparing cachectic and non-cachectic patients, peak VO_2, LVEF, treatment, and mean functional NYHA class (2.6±0.1 vs 2.8±0.1) were not found to be significantly different (all p>0.2). In all subjects blood samples were collected from an antecubital polyethylene catheter in the morning, between 9 and 10 AM, after a fasting period of _12 hours with supine rest of at least 20 min.

Results. The main results are shown in Table 1. When the CHF patients were subdivided according to presence or absence of cachexia, major differences in the measures of the GH - IGF axis were found. Non-cachectic CHF patients and controls did not differ.

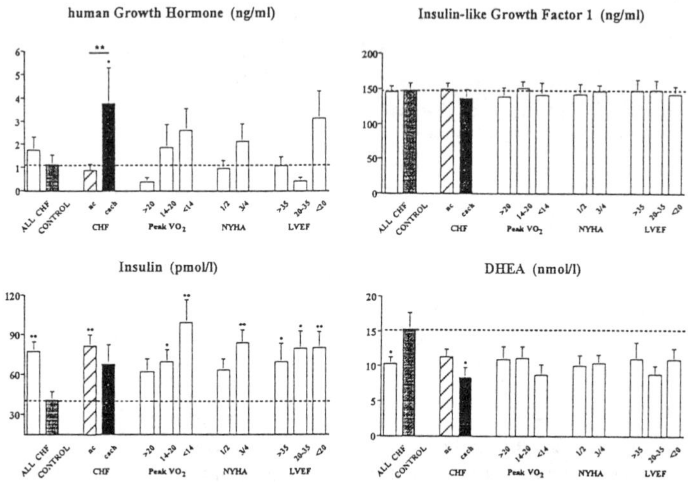

Figure 1. Levels of anabolic hormones (hGH, IGF-1, Insulin, DHEA) in 16 healthy controls and 53 patients with chronic heart failure (CHF). Data are presented as mean±SEM.

- cachectic state: nc=non-cachectic (n=37); cach=cachectic (n=16)
- maximal oxygen consumption [(peakVO^2)<14 (n=17) vs 14-20 (n=24) vs > 20 mL/kg/min (n=12)]
- functional New York Heart Association class (NYHA), class1-2 (n=16) vs class 3-4 (n=37)
- left ventricular ejection fraction (LVEF), < 20 (n=24) vs 20-35 (n=17) vs > 35 (n=12)

*p<0.05 and **p<0.01 for inter-group comparision; •p<0.05 and ••p<0.01 vs controls; *taken from reference 6, with permission*

Table 1. Parameters of the GH-IGF1 axis in healthy controls and in patients with chronic heart failure (CHF). Cachectic CHF subjects had an history of non-intentional weight loss > 7.5% compared to previous normal weight, whereas the non-cachectic CHF patients had no history of weight loss. +p<0.1, *p<0.05, $p<0.01, £p<0.001, çp<0.0001 vs controls/non-cachectic CHF patients, respectively. *ANOVA p=0.13.

	Controls	All CHF patients	Non-cachectic CHF	Cachectic CHF
Total GH (ng/mL) (n=26/72/51/21)	1.22±0.31	2.4±0.47	1.2±0.28	5.31±1.29 (ç/ç)
Intact GH (ng/mL) (n=24/59/38/21)	0.53±0.12	1.05±0.2	0.58±0.14	1.91±0.47 (£/£)
IGF-1 (ng/mL) (n=26/72/51/21)	151±9	142±6	150±8	124±9 (+/+)
IGF-BP3 (ng/mL) (n=24/57/36/21)	3.75±0.14	3.49±0.12	3.7±0.13	3.13±0.21 (*/*)
GH-BP (pmol/L) (n=24/57/36/21)	852±83	823±62	950±84	607±64 (+/$)
Log (IGF1/GH) (n=26/72/51/21)	2.72±0.19	2.47±0.12	2.8±0.14	1.68±0.16 (£/ç)

The cachectic CHF patients as a group presented the biochemical characteristics of the syndrome of acquired GH resistance. Compared to non-cachectic CHF patients they had increased levels of total GH (+342%, ANOVA $p<0.0001$), and immunofunctionally intact GH (+229%, $p=0.0002$). Compared to non-cachectic patients the cachectic patients showed a trend towards lower IGF-I (-17%, $p=0.06$), a significantly reduced IGFBP3 (-15%, $p=0.01$), and a reduction of GHBP (-36%, $p=0.005$). Consequently, the ratio of IGF-I to total GH (Nichols) was markedly reduced in cachectic CHF patients (log IGF-I/GH in cachectics 1.68±0.16, i.e. -40% compared to non-cachectics ($p<0.0001$) and -38% compared to controls ($p=0.0002$)) (Figure 2).

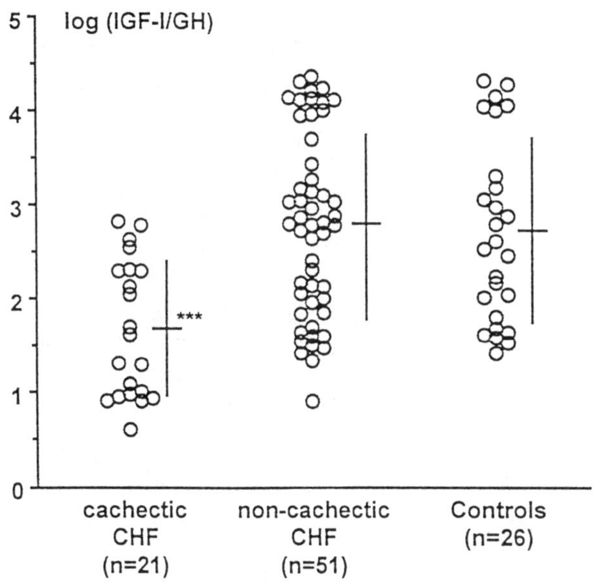

Figure 2. The relation between circulating GH and IGF-1 levels [log (IGF-1/GH) ratio] in patients with chronic heart failure (CHF) and healthy control subjects. Mean values and standard deviations are marked. ***$p<0.0001$ vs non-cachectic CHF and $p=0.0001$ vs controls.

As there was considerable overlap of the IGF-I/GH ratio in cachectic and non-cachectic patients we have also analyzed the clinical characteristics of patients in the lowest tertile of IGF-I/GH ratios (n=24) vs the remaining 48 CHF patients. This analysis showed that patients with low IGF-I/GH ratio were more frequently cachectic (50% vs 19%), that they had a lower peak VO_2 (both $p=0.006$), and that they had lower GHBP ($p=0.003$) and IGFBP3 levels ($p<0.0001$). In contrast, LVEF, NYHA class, and CHF aetiology were similar. Additionally, we assessed the characteristics of the 12 non-cachectic patients with low IGF-I/GH ratio (from the group of patients in the overall lowest tertile in the first comparison) and compared these patients with the features of the 39 non-cachectic patients with normal or high IGF-I/GH ratios. Non-cachectic patients with low IGF-I/GH ratio had raised GH levels, but IGF-I, GHBP, and IGFBP3 were not significantly altered (all $p>0.19$). Nevertheless, these non-cachectic patients tended to have lower BMI ($p=0.08$) and % ideal weight ($p<0.06$), and they showed signs of sympathetic activation (higher noradrenaline and adrenaline levels, both $p<0.002$). Again, LVEF, NYHA class, and CHF aetiology were similar ($p>0.20$, data not shown).

Correlation Analyses. The relationship of the log IGF-I/GH ratio with clinical characteristics and IGFBP3 and GHBP are detailed in Table 2.

Table 2. The relation of the biochemical marker of GH resistance log (IGF-1/GH) to clinical and biochemical parameters in healthy controls and in patients with with chronic heart failure (CHF). Regression coefficients of simple linear regression are given: *p<0.05, **p<0.01, $p<0.001, £p<0.0001.

	All subjects	Controls	All CHF patients	Non-cachectic CHF	Cachectic CHF
Age	-0.41£	-0.26	-0.45£	-0.35*	-0.26
BMI	0.48£	0.23	0.54£	0.37**	0.22
% weight	0.48£	0.23	0.54£	0.37**	0.20
LVEF	-	-	0.04	-0.06	0.36
NYHA class	-	-	-0.20	-0.11	-0.49*
Peak VO_2	0.21*	-0.07	0.35**	0.28*	0.48*
IGF-BP3	0.36**	0.06	0.43$	0.24	0.52*
GHBP	0.65£	0.79£	0.61£	0.50**	0.68$

The strongest predictor of the log IGF-I/GH ratio was the GHBP plasma concentration (control subjects: n=24, r=0.79, $p<0.0001$; all CHF: n=57, r=0.61, $p<0.0001$; non-cachectic CHF: n=36, r=0.50, p=0.002; cachectic CHF: n=21, r=0.68, p=0.0007). In cachectic patients total weight loss (in % of previous weight) correlated also with the GHBP levels (r=-0.60, p=0.004). In non-cachectic patients no parameter could predict lower GH sensitivity independently of the GHBP levels. In cachectic patients the GHBP level were related to log IGF-I/GH ($p<0.05$) independently of age, peak VO_2, LVEF, and the amount of previous weight loss in % (multivariate analysis). In the cachectic CHF patients we found a close relationship between GHBP and IGFBP3 (r=0.70, p=0.0005). In non-cachectic patients (r=0.28, p=0.10) and controls (r=0.16, p=0.44) this relationship was not present.

In a subgroup of 46 CHF patients we were able to adjust the correlation analysis for body composition as determined by dual energy X-ray absorptiometry (DEXA) scanning. GHBP levels related to lean tissue content (r=0.44, $p<0.003$) and fat tissue content (r=0.56, $p<0.0001$). A multivariate analysis of the relationship between GHBP vs fat tissue content and log IGF-I/GH in these patients showed that GHBP was related to the log IGF-I/GH ratio (standardized coefficient = 0.52, p=0.0001) independently of its correlation to fat tissue content (standardized coefficient = 0.31, p=0.014; adjusted joint $r^2=0.49$, $p<0.0001$).

ACQUIRED GH RESISTANCE IN VARDIAC CACHEXIA

Our studies indicate that the presence of cachexia in patients with CHF is linked to the biochemical pattern of acquired GH resistance. The principal predictor of the biochemical features of GH resistance appear to be GHBP plasma levels. A reduction of tissue GH receptors thus may be causal for the presence of GH resistance.

GH is secreted in a pulsatile fashion, and this may make analyses using single morning blood samples questionable. However, it is now emerging that in contrast to healthy subjects in CHF patients a strong relationship between results of analyses of overnight blood sampling and single morning samples can be found (Maurizio Volterrani, personal communication). The measure of GH sensitivity (log IGF-I/GH) was on average 1.12 units lower in cachectic patients compared to non-cachectic patients with CHF (-40%, Table 2), but it is important to note that the true ratio (before logarithmic transformation) was more than 10-fold lower in the cachectic patients. Nevertheless, there was overlap between groups for the IGF-I/GH ratio (Figure 2), which implies that the biochemical features of GH resistance are not completely restricted to cachectic patients (12 of 51, i.e. 24%, of the non-cachectic patients had a log IGF-I/GH ratio in the lowest tertile of the CHF cohort). From our studies we would estimate that approximately 60 to 70% of cachectic CHF patients and in 20 to 30% of non-cachectic CHF patients are GH resistant.

The main predictor of a lower IGF-I/GH ratio was the level of GHBP. As GHBP levels reflect the cellular GH receptor status (23,24), the development of GH resistance may be due to changes at the cellular level. These changes occur independently of age and clinical status, therefore, the presence of GH resistance cannot be predicted from conventional markers of CHF severity. A reduction of the fat tissue content in cachectic patients could *per se* contribute to a reduction of GHBP (28,29) leading to a reduced IGF-I/GH ratio. However, in our investigations the relationship between GHBP and the IGF-I/GH ratio was much stronger and independent of the fat tissue content as assessed by DEXA scanning.

Similar to resistance to the action of insulin (30), GH resistance may be a potentially important hormonal resistance syndrome in CHF. Particularly in non-cachectic patients with low IGF/GH ratio (in study 1) it was interesting to note that these patients showed signs of sympathetic activation and lowered BMI, although they have not yet developed weight loss as a consequence. These non-cachectic patients may be metabolically similar to cachectic patients. We have previously shown (6) that only a small proportion of non-cachectic patients have abnormal neurohormonal status. Potentially, the non-cachectic patients with low IGF/GH ratio are those who are at a higher risk of developing cachexia and may be (already) GH resistant.

We could not demonstrate significant reductions in IGF-I levels, but trends towards deminished IGF-I levels in cachectic compared to non-cachectic CHF patients were found. We are not able to comment on IGF-I tissue levels in cachectic CHF patients. However, in a recent report on tissue IGF-I mRNA expression in skeletal muscle tissue of non-cachectic patients, Hambrechts group found tissue IGF-I mRNA to be significantly reduced (31). Low tissue expression of IGF-I could be important for the development of the wasting process. IGF-I can induce pleiotropic responses in many cell types, including the regulation of growth and differentiation. It has been demonstrated by many authors that IGF-I promotes tumour growth (32-35), probably by suppressing apoptosis (36-41). Lack of IGF-I may favour the development of muscle wasting in the presence of sympathetic activation and catabolic stimuli (42).

GH TREATMENT IN CHF AND GH RESISTANCE

In 1996 a pilot study of 7 CHF patients showed that cardiac function and exercise tolerance improved after 3 months treatment with GH (14 IU per week) (43). Frustaci et al (44) were not able to document any significant benefit of 3 months treatment with GH (n=5, dose: 28 IU per week). Recent randomized controlled studies investigating the effects of GH in a total of 72 CHF patients (2 IU/d) (45-46) have shown that GH treatment is safe, but it did not improve NYHA functional class, LVEF, 6 min walking time, or neurohormonal status. In the larger study [n=50, (45)] the only significant non-hormonal effect of GH treatment was an increase of the left ventricular mass as assessed by magnetic resonance imaging. Previously, haemodynamic benefits of 24 hours intravenous GH infusion have been reported (47), but in randomized studies (45) this could not be confirmed. Recently, Genth-Zotz et al. (48) reported that open GH treatment (2 IU/d) over a period of 3 months significantly improved haemodynamics, clinical status, and exercise capacity in 7 CHF patients with an ischaemic aetiology. If the differences between the studies are not entirely due to the placebo effect of GH treatment, the reasons for differing results must rest with the patient groups studied as the treatment scheme was similar in all studies. Conventional markers of disease severity were also similar in the different trials. Therefore, the observed differences may be due to metabolic differences between patients that may exist but were not assessed. We have demonstrated that, in particular, cachectic patients but also some non-cachectic CHF patients show the characteristics of acquired GH resistance and this may prevent a beneficial response to GH therapy at a fixed dose. In the future it may be considered beneficial to recruit patients for GH treatment on the basis of a first dose IGF-I response or according to GHBP baseline plasma levels. Future studies would need to prospectively determine useful cut-off values.

THERAPEUTIC IMPLICATIONS

At present, there is no specific treatment available for patients with cardiac cachexia. New treatments need to be developed, and these should be based on insights into the underlying pathophysiology. Overcoming acquired growth hormone resistance may be a way to go. Theoretically, there are several possibilities to overcome GH resistance in CHF patients. Firstly, the therapeutic dose of GH could be increased, although this is generally thought to be risky due to the many possible side effects (promotion of a diabetogenic condition, sympathetic activation (49) or sodium/water retention and carpal tunnel syndrome (50). Most recently, a study in critically ill adults who had been in an intensive care unit for 5 to 7 days (due to cardiac surgery, abdominal surgery, multiple trauma, or acute respiratory failure) reported the results of GH treatment using 5.3 (16 IU) or 8.0 mg (24 IU) GH per day compared to placebo (51). In this study, the mortality in the GH treatment group was about 2-fold higher than in the placebo group. The majority of the deaths was attributable to multiple organ failure and septic shock or uncontrolled infection, and >50% of the deaths occurred in the first 10 days of treatment. It has been suggested that the outcome of these patients after GH treatment was poor because these patients very likely did not have systemic inflammation and neurohormonal activation as a component of their disease (52). Amongst CHF patients, cachectic patients clearly show the highest plasma levels of catecholamines and inflammatory cytokines (6,9).

In the ICU trial (51) it was not the focus to achieve anabolic effects on lean tissue, however, this would be the aim of treating cachectic CHF patients. The metabolic status of cachectic CHF patients may be somewhat comparable to that of patients with AIDS-associated wasting, and these patients also show the biochemical characteristics of GH resistance. In the USA higher dose GH therapy has been approved by the Food and Drug Administration for use in treating AIDS-associated wasting (4 to 6 mg, i.e. 12 to 18 IU, to be administered

by subcutaneous injection daily) (53). The most common side effects are increases of tissue turgor complex and musculoskeletal discomfort (54). The experience with anabolic doses of GH in CHF patients is very limited. Two case reports (55,56), however, demonstrated that short periods (1 week to 3 months) of high-dose GH therapy (70 to 98 IU per week) of altogether 3 cachectic patients with CHF resulted in profound increases of muscle mass and strength and improvement of exercise capacity with no reported side effects.
Alternatively, IGF-I itself might be given to overcome acquired growth hormone resistance and to achieve anabolic effects. It is known that IGF-I can reverse states of insulin resistance independently of insulin (57,58) and can ameliorate weight loss induced by starvation (59). IGF-I can induce hypoglycemia (the acute effectiveness of IGF compared to insulin in healthy adults is about 6% (60) and also peripheral edema (50). This treatment approach might be considered problematic, as high IGF-I levels were claimed to be associated with increased risk of breast cancer development in premenopausal women less than 50 years old (but not in postmenopausal women) (61), and with the risk of prostate cancer in men (62). Nevertheless, these studies have not assessed the relation between IGF-I and survival, which is important, as high IGF-I levels are also linked to better nutritional status (63) and improved survival in patients with severe liver disease (64). Living longer, could per se increase the chance of developing cancer over time. Interestingly, a study in dogs with heart failure has shown that high IGF-I levels relate to better survival (65). The combination of GH and IGF-I (2,66) may have a particularly powerful anabolic effect, overcome GH resistance, and minimize side-effects. Based on the results of these studies, we propose to perform treatment studies in patients with cardiac cachexia similar to that in cachectic patients with AIDS. The GH doses we would aim to use would be about 50% below the doses used in the ICU trial (51).

CONCLUSIONS

Cardiac cachexia contributes substantially to heart failure related mortality, and can easily be detected. There is still limited understanding of the pathophysiology of body wasting in CHF patients. We found that cachectic patients with CHF show features of acquired GH resistance. A small proportion of non-cachectic patients also show the features of GH resistance. GH resistance may be mediated by a reduction of GH receptors and contribute causally to the syndrome of cardiac cachexia via a reduction of tissue IGF-I. GH resistance might be a major determinant of wasting through different mechanisms: increased catabolism, impaired anabolism and enhanced apoptosis in peripheral tissues. The presence of acquired GH resistance in some patients might be responsible for the differnces between studies using GH in CHF patients. In patients in whom GH treatment is planned, the GH - IGF-I axis should be evaluated before administrating GH over longer periods. Acquired GH resistance could be overcome using higher doses of GH or combining GH with IGF-I.

REFERENCES

1. Katz, AM, Katz, PB. Diseases of heart in works of Hippocrates. Br Heart J 1962;24:257-264.
2. Bentham J, Rodriguez-Arnao J, Ross RJM. Acquired growth hormone resistance in patients with hypercatabolism. Horm Res 1993;40:87-91.

3. Ross RJM, Chew SL. Acquired growth hormone resistance. Eur J Endocrinol 1995;132:655-660.
4. Amselem S, Duquesnoy P, Duriez B, et al. Spectrum of growth hormone receptor mutations and associated haplotypes in Laron syndrome. Hum Mol Genet 1993;2:355-359.
5. Packer M. The neurohormonal hypothesis: a theory to explain the mechanism of disease progression in heart failure. J Am Coll Cardiol 1992;20:248-54.
6. Anker SD, ChuaTP, Ponikowski P, et al. Hormonal changes and catabolic/anabolic imbalance in chronic heart failure and their importance for cardiac cachexia. Circulation 1997;96:526-534.
7. Anker SD, Coats AJS. Cardiac cachexia: a syndrome with impaired survival and immune and neuroendocrine activation. Chest 1999;115:836-847.
8. Anker SD, Ponikowski P, Varney S, et al. Wasting as independent risk factor for mortality in chronic heart failure. Lancet 1997;349:1050-53.
9. Anker SD, Ponikowski PP, Clark AL, et al. Cytokines and neurohormones relating to body composition alterations in the wasting syndrome of chronic heart failure. Eur Heart J 1999;20:683-693.
10. Anker SD, Clark AL, Teixeira MM, Hellewell PG, Coats AJS. Loss of bone mineral in patients with cachexia due to chronic heart failure. Am J Cardiol 1999;83:612-615.
11. Poehlman ET, Scheffers J, Gottlieb SS, Fisher ML, Vaitekevicius P. Increased resting metabolic rate in patients with congestive heart failure. Ann Intern Med 1994;121:860-862.
12. Pittman JG, Cohen P. The pathogenesis of cardiac cachexia. N Engl J Med 1964;271:403-409.
13. King D, Smith ML, Lye M. Gastro-intestinal protein loss in elderly patients with cardiac cachexia. Age Aging 1996;25:221-223.
14. Anker SD, Chua TP, Swan JW, et al. Hormonal changes and catabolic/anabolic imbalance in chronic heart failure: The importance for cardiac cachexia. Circulation 1997;96:526-534.
15. Mancini DM, Walter G, Reichek N, et al. Contribution of skeletal muscle atrophy to exercise intolerance and altered muscle metabolism in heart failure. Circulation 1992;85:1364-1373.
16. Vescovo G, Serafini F, Facchin L, et al. Specific changes in skeletal muscle myosin heavy chains composition in cardiac failure: differences compared with disuse atrophy as assessed on microbiopsies by high resolution electrophoresis. Heart 1996; 76:337-343.
17. Anker SD. Catecholamine levels and treatment in chronic heart failure. Eur Heart J 1998;19:F56-F61.
18. Ferrari R, Bachetti T, Confortini R, et al. Tumor necrosis factor soluble receptors in patients with various degrees of congestive heart failure. Circulation 1995;92:1479-1486.
19. Levine B, Kalman J, Mayer L, et al. Elevated circulating levels of tumour necrosis factor in severe chronic heart failure. N Engl J Med 1990;323:236-41.
20. Mc Murray J, Abdullah I, Dargie HJ, Saphiro D. Increased concentrations of tumour necrosis factor in cachectic patients with severe chronic heart failure. Br Heart J 1991;66:356-8.

21. Anand IS, Ferrari R, Kalra GS, Wahi PL, Poole-Wilson PA, Harris PC. Edema of cardiac origin. Studies on body water and sodium, renal function, hemodynamic indexes, and plasma hormones in untreated congestive cardiac failure. Circulation 1989;80:299-305.
22. Anker SD, Volterrani M, Pflaum CD, Poole-Wilson PA, Strasburger CJ, Voats AJS. Growth hormone resistance in chronic heart failure. J Am Coll Cardiol 1998;31:204A.
23. Goddard AD, Covello R, Luoh S-M, et al., for the Growth Hormone Insensitivity Study Group. Mutations of the growth hormone receptor in children with ideopathic short stature. N Engl J Med 1995;333:1093-1098.
24. Sotiropoulos A, Goujon L, Simonin G, Kelly PA, Postel-Vinay MC, Finidori J. Evidence for generation of the growth hormone-binding protein through proteolysis of the growth hormone membrane receptor. Endocrinology 1993;132:1863-1865.
25. Pflaum C-D, Dressendörfer RA, Strasburger CJ. A nonisotopic solid phase immunoassay for the determination of human growth hormone binding protein (hGHBP). Exp Clin Endocrinol 1993; 101:44.
26. Strasburger CJ, Wu Z, Pflaum CD, Dressendorfer RA. Immunofunctional assay of human growth hormone (hGH) in serum: a possible consensus for quantitative hGH measurement. J Clin Endocrinol Metab 1996; 81:2613-20.
27. Strasburger CJ, Dressendörfer RA, Lee PD. Non-isotopic two-site immunoassay for IGFBP-3. Growth Regulation 1994;4:138.
28. Fisker S, Vahl N, Jorgensen JO, Christiansen JS, Orskov H. Abdominal fat determines growth hormone-binding protein levels in healthy nonobese adults. J Clin Endocrinol Metab 1997;82:123-128.
29. Roelen CA, Koppeschaar HP, De Vries WR, et al. Visceral adipose tissue is associated with circulating high affinity growth hormone-binding protein. J Clin Endocrinol Metab 1997;82:760-764.
30. Swan JW, Anker SD, Walton C, et al. Insulin resistance in chronic heart failure: Relationship to severity and etiology of heart failure. J Am Coll Cardiol 1997;30:527-532.
31. Schulze PC, Adams V, Linke A, Gielen S, Hambrecht R, Schuler G. Decreased expression of insulin-like growth factor mRNA in muscle biopsies from patients with chronic heart failure. Europ Heart J 1999;20:19.
32. Kaleko M, Rutter WG, Miller AD. Overexpression of the human insulin-like growth factor-I receptor promotes ligand-dependent neoplastic transformation. Mol Cell Biol 1990;10:464-473.
33. Kalebic T, Tsokos M, Helman LJ. In vivo treatment with antibody against the IGF-I receptor suppresses growth of human rhabdomyosarcoma and down-regulates p34/cdc2. Cancer Res 1994;54:5531-5534.
34. Baserga R, Hongo A, Rubini M, Prisco M, Valentinis B. The IGF-I receptor in cell growth, transormation and apoptosis. Biochimica and Biophysica Acta 1997;1332:F105-F126.
35. Arteaga CL, Osborne CK. Growth inhibition of human breast cancer cells in vitro with antibody against the type I somatomedin receptor. Cancer Res 1989;49:6237-6241.
36. Harrington EA, Bennett MR, Fanidi A, Evan GI. C-myc-induced apoptosis in fibroblasts is inhibited by specific cytokines. Embo J 1994;13:3286-3295.
37. Sell C, Baserga R, Rubin R. Insulin-like growth factor I (IGF-I) and the IGF-I receptor prevent etoposide-induced apoptosis. Cancer Res 1995;55:2463-2469.

38. D'Mello SR, Galli C, Ciotti T, Calissano P. Induction of apoptosis in cerebellar granule neurons by low potassium: inhibition of death by insulin-like growth factor-I and cAMP. Proc Natl Acad Sci 1993;90:10989-10993.
39. Singleton JR, Randolph AE, Feldman EL. Insulin-like growth factor-I receptor prevents apoptosis and enhances neuroblastoma tumorigenesis. Cancer Res 1996;56:4522-4529.
40. Resnicoff M, Burgaud JL, Rotman HL, Abraham D, Baserga R. Correlation between apoptosis, tumorigenesis, and levels of insulin-like factor-I receptors. Cancer Res 1995;55:3739-3741.
41. Resnicoff M, Abraham D, Yutanawiboonchai W, et al. The insulin-like growth factor-I receptor protects tumor cells from apoptosis in vivo. Cancer Res 1995;55:2463-2469.
42. Niebauer J, Pflaum C-D, Clark AL, et al. Deficient insulin-like growth factor-1 in chronic heart failure predicts altered body composition, cytokine and neurohormonal activation. J Am Coll Cardiol 1998;32:393-397.
43. Fazio S, Sabatini D, Capaldo B, et al. A preliminary study of growth hormone in the treatment of dilated cardiomyopathy. N Engl J Med 1996;334:809-814.
44. Frustaci A, Gentiloni N, Russo MA. Growth hormone in the treatment of dilated cardiomyopathy. N Engl J Med 1996;335:672-673.
45. Osterziel KJ, Strohm O, Schuler J, et al. Randomised, double-blind, placebo-controlled trial of human recombinant growth hormone in patients with chronic heart failure due to dilated cardiomyopathy. Lancet 1998;351:1233-1237.
46. Isgaard J, Bergh CH, Caidahl K, Lomsky M, Hjalmarson A, Bengtsson BA. A placebo controlled study of growth hormone in patients with congestive heart failure. Eur Heart J 1998;19:1704-1711.
47. Volterrani M, Desenzani P, Lorusso R, D'Aloia A, Manelli F, Giustina A. Haemodynamic effects of intravenous growth hormone in congestive heart failure. Lancet 1997;49:1067-1068.
48. Genth-Zotz S, Zotz R, Geil S, Voigtlander T, Meyer J, Darius H. Recombinant growth hormone therapy in patients with ischemic cardiomyopathy. Effects on hemodynamics, left ventricular function, and cardiopulmonary exercise capacity. Circulation 1999;99:18-21.
49. Loh E, Swain JL. Growth hormone for heart failure - cause for cautious optimism. N Engl J Med 1996;334:856-857.
50. Moxley RT. Potential for growth factor treatment of muscle disease. Curr Opin Neurol 1994;7:427-434.
51. Takala J, Ruokonen E, Webster NR, et al. Increased mortality associated with growth hormone treatment in critically ill adults. N Engl J Med 1999;341:785-792.
52. Demling R. Growth hormone therapy in critically ill patients. N Engl J Med 1999;341:837-839.
53. Windisch PA, Papatheofanis FJ, Matuszewski KA. Recombinant human growth hormone for AIDS-associated wasting. Ann Pharmacother 1998;32:437-445.
54. Van Loon K. Safety of high doses of recombinant human growth hormone. Horm Res 1998;49:78-81.
55. Cuneo RC, Wilmshurst, P, Lowy, McGauley G, Sönksen PH. Cardiac failure responding to growth hormone. Lancet 1989;333:838-839.
56. O'Driscoll JG, Green DJ, Ireland M, Kerr D, Larbalestier. Treatment of end-stage cardiac failure with growth hormone. Lancet 1997;349:1068.

57. Usala A-L, Madigan T, Burguera B, et al. Brief report: treatment of insulin-resistant diabetic ketoacidosis with insulin-like growth factor 1 in an aldolescent with insulin-dependent diabetes. N Engl J Med 1992;327:853-857.
58. Dimitriadis G, Parry-Billings M, Bevan S, et al. Effects of insulin-like growth factor 1 on the rates of glucose transport and utilization in rat skeltal muscle in vitro. Biochem J 1992;285:269-275.
59. O'Sullivan U, Gluckman PD, Breier BH, Woodall S, Siddiqui RA, McCutcheon SN. Insulin-like growth factor 1 (IGF-1) in mice reduces weight loss during starvation. Endocrinology 1989;125: 2793-2794.
60. Guler H-P, Zapf J, Froesch ER. Short-term metabolic effects of recombinant human insulin-like growth factor in healthy adults. N Engl J Med 1987;317:137-140.
61. Hankinson SE, Willett WC, Colditz GA, et al. Circulating concentrations of insulin-like growth factor-I and risk of breast cancer. Lancet 1998;351:1393-1396.
62. Chan JM, Stampfer MJ, Giovannucci E, et al. Plasma insulin-like growth factor-I and prostate cancer risk: a prospective study. Science 1998;279:563-566.
63. Qureshi AR, Alvestrand A, Danielsson A, et al. Factors predicting malnutrition in hemodialysis patients: a cross-sectional study. Kidney Int 1998;53:773-782.
64. Moller S, Becker U, Juul A, Skakkebaek NE, Christensen E. Prognostic value of insulinlike growth factor I and its binding protein in patients with alcohol-induced liver disease. Hepatology 1996;23:1073-1078.
65. Freeman LM, Rush JE, Kehayias JJ, et al. Nutritional alterations and the effect of fish oil supplementation in dogs with heart failure. Vet Intern Med 1998;12:440-448.
66. Jenkins RC, Ross RJM. Growth hormone therapy for protein catabolism. Q J Med 1996;89:813-819.

9
GROWTH HORMONE TREATMENT IN EXPERIMENTAL HEART FAILURE

Jörgen Isgaard and Bengt-Åke Bengtsson
Research Center for Endocrinology and Metabolism, Sahlgrenska University Hospital, Göteborg, Sweden

INTRODUCTION

During the last years, GH and IGF-I have been recognized as important regulators of cardiac structure and function. The effects of GH on peripheral tissues are thought to be either direct or mediated by IGF-I. IGF-I is synthesized in multiple tissues including the heart and some of the effects have been suggested to be paracrine or autocrine (1-4). Both GH and IGF-I receptors are expressed in the heart and provide possibilities for GH and IGF-I signalling. Importance of the GH/IGF-I axis for the development of cardiac hypertrophy in response to increased hemodynamic load will be discussed in this chapter along with an overview of GH and IGF-I effects on intracellular calcium handling, apoptosis and cardiac performance in various experimental models.

GH/IGF-I AXIS AND CARDIAC HYPERTROPHY

IGF-I

Several studies have demonstrated an increased left ventricular (LV) expression of IGF-I in different rat models of hypertension. The induction of IGF-I expression preceded or occurred in parallel with development of LV hypertrophy (4,5,6). These results suggest that hypertension and increased LV pressure load provide a mechanical stimulus for a local growth response in the challenged ventricle where locally synthesized IGF-I may play a role through autocrine/paracrine mechanisms. Results supporting the hypothesis that an increased wall stress may trigger a local induction of IGF-I expression have also been reported in experimental models where the RV is predominantly challenged either through a surgically induced fistula between the abdominal aorta and the caval vein (ACF) resulting in volume overload (7) or through an induction of pulmonary hypertension (8).

GH receptor

The GH-receptor (GH-R) gene is expressed in the rat (9) and rabbit (10) heart. Basal levels of cardiac GH-R are lower than in the liver but the GH-R expression in the heart appears to be relatively high compared to other tissues (10).

GH-R gene expression was shown to be elevated in the acutely overloaded RV of ACF-operated rats (9) and in the LV of 2K1C rats (6), suggesting a role for the GH-R in compensatory growth of a hemodynamically challenged heart. However, since induction of GH-R expression preceded IGF-I mRNA in the volume overload model (9) but not in the

2K1C rats (6), it is unclear whether increased GH-R expression is really required for the upregulation of IGF-I message. Alternatively, it may be speculated that an increased expression of GH-R sustains the local synthesis of IGF-I in the heart. In summary, an upregulation of both IGF-I and GH-R expression is linked to increased hemodynamic load and suggests a potential role for the GH-IGF-I axis during development of compensatory cardiac hypertrophy.
The hearts of patients with idiopathic dilated cardiomyopathy are characterized by ventricular dilation, thin ventricular walls and increased wall stress. A possible increase of ventricular mass through growth promoting effects of GH and/or IGF-I with a subsequent reduction of wall stress by GH treatment in these patients would be beneficial and this was observed in the study by Fazio and co-workers (11).

CARDIAC EFFECTS OF GH AND IGF-I

Although components of the GH/IGF-I axis appear to be linked to development of cardiac hypertrophy, the precise mechanisms of actions of GH and IGF-I on the heart are mostly unknown and there are few *in vitro* studies concerning GH/IGF-I effects on cardiomyocytes. In papillary muscle from rats exposed to very high GH plasma levels due to a GH secreting tumor, it was found that the maximum Ca^{2+} activated force per cross-sectional area was increased, suggesting a GH induced elevated Ca^{2+} responsiveness of the myofilaments (12). In a model with isolated buffer-perfused hearts from rats treated with high doses of GH or IGF-I for four weeks, systolic function was improved and findings supported the notion of an increased maximal response to Ca^{2+} (13). In a recent study, rats with postinfarction heart failure treated with GH for 14 days showed enhanced peak systolic intracellular Ca^{2+} transients and increased sarcoplasmic reticulum Ca^{2+} ATPase 2 with resulting improvement of contractile reserve compared to controls (14). However, in this study, no beneficial effects of GH on *in vivo* LV function were found which differs from other studies. In addition to direct effects on the heart, a lowering of systemic vascular resistance by GH may also be beneficial for hemodynamics (15). It may be speculated that this effect is rapid and could be an important factor in studies demonstrating acute cardiovascular effects of GH.
In vitro studies, with addition of IGF-I to cultured neonatal rat cardiomyocytes, have demonstrated changes associated with hypertrophy including increments in cell size, protein synthesis and induced expression of myosin light chain-2 and troponin I (16). In adult rat cardiomyocytes, it has been reported that IGF-I stimulates myofibril development (17) and increases isometric force and free cytosolic Ca^{2+} (18).
A number of experimental studies both in normal rats (19) and in rats with surgically induced myocardial infarction (20-22) show that GH and IGF-I, alone or in combination, improve cardiac function. Moreover, GH and IGF-I were still effective in improving cardiac function when added as an adjunct treatment to post-infarction rats receiving ACE inhibition (15). This is an important finding since ACE inhibition is standard treatment for patients with congestive heart failure. However, at present the interactions between GH and the renin-angiotensin system are not fully clarified. Most of the effects of GH and IGF-I on heart function that have been reported are related to systolic function and include increased stroke volume and cardiac index measured invasively (15,20,21) or by echocardiography (22). Hypertrophy of the heart is also a common observation, although this effect may be partly time- and dose dependent since beneficial effects of GH on systolic function have been reported also in the absence of hypertrophy (22). However, the absence of positive effects of GH on the impaired cardiac function after experimental myocardial infarction (23) and ventricular pacing of dogs (24) has also been reported. Interestingly, it was recently demonstrated that GH, apart from improving

cardiac function, also attenuates left ventricular remodeling in rats after myocardial infarction, apparently mostly due to reduction of LV dilation and additional hypertrophy of the noninfarcted myocardium (25).

Apoptosis

Apoptosis, or programmed cell death, has been reported to occur in the myocardium of patients both after ischemic injury (26) and in conditions with dilated cardiomyopathy (27) Some of the genes involved in the apoptotic process have been identified including the Fas-receptor which has been shown to mediate signals for apoptosis (28,29). Upregulation of the Fas-receptor has been shown to be associated with apoptosis after myocardial infarction in rats (30). It has been reported that administration of IGF-I and GH prior to cardiac ischemia attenuates the increased rate of myocyte apoptosis (31,32). Using transgenic mice overexpressing IGF-I, the group by Anversa and co-workers found that apoptosis was decreased and cardiac function was preserved compared to non-transgenic littermates when these mice were subjected to experimental myocardial infarction (33). Wang and co-workers recently demonstrated in H9C2 cardiac muscle cells that apoptosis induced by serum deprivation and doxorubicin, was partially attenuated by IGF-I through inhibition of Bax induction, caspase 3 activation and DNA fragmentation (34).

SUMMARY

There is ample evidence to support a role for the GH/IGF-I axis in regulation of cardiac growth, structure and function. GH may act directly on the heart or through circulating IGF-I. Moreover, GH has been found to regulate local production of IGF-I in the heart. Both the GH-R and IGF-I-R are expressed in cardiac tissue. Hence, the IGF-I-R receptor can theoretically be activated through locally produced IGF-I acting via autocrine/paracrine mechanisms, or via circulating IGF-I exerting its effects as an endocrine agent. During conditions of pressure and volume overload, an increased systolic wall stress triggers an induction of gene expression of IGF-I, GH-R and possibly IGF-I-R implying a potential role for the GH/IGF-I axis in the development of adaptive hypertrophy of the heart and vessels. Cardiovascular effects of GH in experimental studies include beneficial effects on systolic function and a decrease of peripheral resistance. Experimental studies suggest an increased Ca^{2+} responsiveness and upregulation of sarcoplasmic reticulum Ca^{2+} ATPase 2 as possible underlying causes for the improvement in contractility, although effects of GH and IGF-I on apoptosis may possibly also play a more long-term role for cardiomyocyte survival. It is evident that further basic studies are required to gain insight into the GH and IGF-I mechanisms of action on the heart if GH is considered as adjuvant therapy for congestive heart failure.

REFERENCES

1. D'Ercole AJ, Stiles AD, Underwood LE. Tissue concentrations of somatomedin-C: Further evidence for multiple sites of synthesis and paracrine orautocrine mechanisms of action. Proc Natl Acad Sci USA 1984;81:935-939.
2. Murphy LJ, Bell GI, Duckworth L, Friesen HG. Identification, characterization and regulation of a rat complementary deoxyribonucleic acid which encodes insulin-like growth factor I. Endocrinology 1987;121:684-691.

3. Isgaard J, Nilsson A, Vikman K, Isaksson OGP. Growth hormone regulates the level of IGF-I mRNA in rat skeletal muscle. J Endocrinol 1989;120:107-112.
4. Wåhlander H, Isgaard J, Jennische E, Friberg P. Left ventricular insulin-like growth factor I mRNA increases in early renal hypertension. Hypertension 1992;19:25-32
5. Donohue TJ, Dworkin LD, Lango MN, et al. Induction of myocardial insulin-like growth factor I gene expression in left ventricular hypertrophy. Circulation 1994;89:799-809.
6. Guron G, Friberg P, Wickman A, Brantsing C, Gabrielsson B, Isgaard J. Cardiac insulin-like growth factor I and growth hormone receptor expression in renal hypertension. Hypertension 1996;27:636-642.
7. Isgaard J, Wåhlander H, Adams MA, Friberg P. Increased expression of growth hormone receptor mRNA and insulin-like growth factor I mRNA in volume overloaded hearts. Hypertension 1994;23:884-888.
8. Russell-Jones DL, Leach RM, Ward JPT, ThomasCR. Insulin-like growth factor-I gene expressionis increased in the right ventricle hypertrophy induced by chronic hypoxia in the rat. J Endocrinol. 1993;10:99-102.
9. Mathews LS, Engberg B, Norstedt G. Regulation of rat GH receptor gene expression. J Biol Chem 1989;264:9905-9910.
10. Ymer SI, Herington AC. Developmental expression of the growth hormone receptor gene in rabbit tissues. Mol Cell Endocrinol 1992;83:39-49.
11. Fazio S, Sabatini D, Capaldo B, et al. A preliminary study of growth hormone in the treatment of dilated cardiomyopathy. N Engl J Med 1996;334:809-814.
12. Mayoux E, Ventura-Clapier R, Timsit J, Behar-Cohen F, Hoffman C, Mercadier JJ. Mechanical properties of rat cardiac skinned fibers are altered by chronic growth hormone hypersecretion. Circ Res 1993;72:57-64.
13. Strömer H, Cittadini A, Douglas PS, Morgan JP. Exogenously administered growth hormone and insulin-like growth factor-I alter intracellular Ca^{2+} handling and enhance cardiac performance. In vitro evaluation in the isolated isovolumic buffer-perfused rat heart. Circ Res 1996;79:227-236.
14. Tajima M, Weinberg EO, Bartunek J, et al. Treatment with growth hormone enhances contractile reserve and intracellular calcium transients in myocytes from rats with postinfarction heart failure. Circulation 1999;127-134.
15. Jin H, Yang R, Gillett N, Clarck RG, Ko A, Paoni NF. Beneficial effects of growth hormone and insulin-like factor-1 in experimental heart failure in rats treated with chronic ACE inhibition. J Cardiovasc Pharmacol 1995;26:420-425.
16. Ito H, Hiroe M, Hirata Y, et al. Insulin-like growth factor-I induces hypertrophy with enhanced expression of muscle-specific genes in cultured rat cardiomyocytes. Circulation 1993;87:1715-1721.
17. Donath MY, Zapf J, Eppenberger-Eberhardt M, Froesch ER, Eppenberger HM. Insulin-like growth factor I stimulates myofibril development and decreases smooth muscle I-actin of adult cardiomyocytes. Proc Natl Acad Sci USA 1994;91:1686-1690.
18. Freestone NS, Ribaric S, Mason WT. The effect of insulin-like growth factor-1 on adult rat cardiac contractility. Mol Cell Biochem 1996;163/164:223-229.
19. Cittadini A, Strömer H, Katz SE, et al. Differential cardiac effects of growth hormone and insulin-like growth factor-I in the rat. A combined in vivo and in vitro evaluation. Circulation 1996;93:800-809.
20. Yang R, Bunting S, Gillett N, Clark R, Jin H. Growth hormone improves cardiac performance in experimental performance. Circulation 1995;92:262-267.

21. Duerr RL, Huang S, Miraliakbar HR, Clarck R, Chien KR, Ross JJr. Insulin-like growth factor-1 enhances ventricular hypertrophy and function during the onset of experimental cardiac failure. J Clin Invest 1995;95:619-627.
22. Isgaard J, Kujacic V, Jennische E, et al. Growth hormone improves cardiac function in rats with experimental myocardial infarction. Eur J Clin Invest 1997;27:517-525.
23. Shen YT, Wiedmann RT, Lynch JJ, Grossman W, Johnson RG. GH replacement fails to improve ventricular function in hypophysectomized rats with myocardial infarction. Am J Physiol 1996;271:H1721-H1727.
24. Shen YT, Woltmann RF, Appleby S, et al. Lack of beneficial effects of growth hormone treatment in conscious dogs during development of heart failure. Am J Physiol 1998;274:H456-H466.
25. Cittadini A, Grossman J, Napoli R, et al. Growth hormone attenuates early ventricular remodeling and improves cardiac function in rats with large myocardial infarction. J Am College Cardiol 1997;29:1109-1116.
26. Saraste A, Pulkki K, Kallajoki, M, Henriksen K, Parvinen M, Voipio-Pulkki LM. Apoptosis in human acute myocardial infarction. Circulation 1997;95:320-323.
27. Narula J, Haider N, Virmani R, et al. Apoptosis in myocytes in end-stage heart failure. N Engl J Med 1996;335-1182-1189.
28. Itoh N, Yonehara S, Ishii A, et al. The polypeptide encoded by the cDNA for human cell surface antigen Fas can mediate apoptosis. Cell 1991;66:233-243.
29. Nagata S, Goldstein P. The Fas death factor. Science 1995;267:1449-1456.
30. Kajstura J, Cheng W, Reiss K, et al. Apoptotic and necrotic myocyte cell deaths are independent contributing variables of infarct size in rats. Lab Invest 1996;74:86-107.
31. Buerke M, Mohara T, Skurk C, Nuss C, Tomaselli K, Lefer AM. Cardioprotective effect of insulin-like growth factor-I in myocardial ischemia followed by reperfusion. Proc Natl Acad Sci USA 1995;92:8031-8035.
32. Buerke M, Prufer D, Ibe W, et al. Human growth hormone exerts cardioprotective effects in murine reperfusion injury. Abstract # 2116, 70th Scientific Session of the American Heart Association, Orlando, November 1997
33. Li Q, Li B, Wang X, et al. Overexpression of insulin-like growth factor-1 in mice protects from myocyte death after infarction, attenuating ventricular dilation, wall stress and cardiac hypertrophy. J Clin Invest 1997;100:1991-1999.
34. Wang L, Weiqiong MA, Markovich R, Wen-Lieng L, Wang PH. Insulin-like growth factor I modulates induction of apoptotic signaling in H9C2 cardiac muscle cells. Endocrinology 1998;139:1354-1360.

10
ACUTE GROWTH HORMONE EFFECTS IN HUMAN SUBJECTS WITH HEART FAILURE

Filippo Manelli[1], Maurizio Volterrani[2], Roberto Lorusso[3], Giuseppe Romanelli[1], Andrea Giustina[1]

[1]Endocrine and Geriatric Sections, Dept of Internal Medicine, University of Brescia; [2]Fondazione "S. Maugeri", Cardiology Dept., Gussago Medical Centre, Brescia; [3]II Cardiac Surgery, Spedali Civili di Brescia; Italy

INTRODUCTION

Growth hormone (GH) is involved, either directly or via its biological mediator insulin-like growth factor-1 (IGF-1), in several physiological processes; in fact, a significant role for this hormone in the regulation of lean and fat mass and intermediary metabolism has been demonstrated (1-3). Moreover, it has long been known that GH has structural effects on the left ventricular muscle (4). These structural effects of GH on the heart have been confirmed in the human by data showing that recombinant human GH, administered for three months in patients with idiopathic dilated cardiomyopathy and congestive heart failure (CHF), increases myocardial mass and reduces the volume of left ventricular chamber, resulting in improvement in hemodynamics, myocardial energy metabolism and clinical picture (5). Recently, several independent lines of evidence have suggested that GH may also play an important role in the regulation of heart function. In fact, administration of exogenous GH causes cardiac hypercontractility in normal subjects (6). Moreover, in an experimental rat model of heart failure it has been shown that GH administration enhances cardiac output and stroke volume by increasing myocardial contractility and decreasing peripheral vascular resistances (7). Finally, acute (24h) decrease in circulating GH levels has been shown to cause an improved cardio-pulmonary function in patients with acromegaly (8).
We assessed the functional role of GH in CHF examining the acute effects of a short-term continuous intravenous infusion of recombinant GH on hemodynamic parameters in twelve patients with dilated cardiomyopathy and CHF.

METHODS

Twelve male patients with chronic ischemic or idiopathic CHF (see Table 1 for clinical characteristics) entered the study after a 7-day clinical and therapeutical stabilization period in the hospital. The day before the hemodynamic study, all the patients underwent a two dimensional echocardiography and a maximal symptom-limited bicycle exercise test with expired gases collection. The overall duration of the hemodynamic study was 48 hours over a three-day period. The first 24 hours after right heart catheterization (from 8.00 a.m. of the first day of the study to 8.00 a.m. of the second day) were considered the control period; in fact, during the following 24 hours (from 8.00 a.m. of the second day of the study to 8.00 a.m. of the third day) all the patients underwent intravenous constant pump infusion (Abbott, Lifecare 5000, North Chicago, Illinois, USA) of recombinant human GH (Humatrope, Eli Lilly, Italy). Hemodynamic parameters were monitored during the 48 hours of the study.

Table 1. Clinical characteristics of the patients. EF: ejection fraction, measured by echocardiography; Peak VO_2: VO_2 at peak exercise; NYHA: New York Heart Association Class; F: frusemide; ACE-I: ACE-inhibitors; V: vasodilators; D: digoxin; As: aspirin; Wa: warfarin; Am: amiodarone; Hy: hydralazine.

Patient #	EF (%)	Peak VO2 (mL/kg/min)	NYHA	Etiology	Duration of disease (months)	Treatment
1	20	13	2	ischemic	20	F, ACE-I, V, D, As
2	21	12	4	ischemic	12	F, ACE-I, V, D, Wa
3	19	10	4	ischemic	36	F, ACE-I, V, D, Wa
4	22	16	3	ischemic	24	F, ACE-I, V, D, As, Am
5	36	14	2	ischemic	48	F, ACE-I, V, D, As, Am
6	23	9	3	idiopathic	48	F, ACE-I, V, D, Wa, Am
7	26	7.4	3	ischemic	48	F, ACE-I, V, D
8	24	5.7	4	ischemic	12	F, ACE-I, V, D, Wa, Am
9	13	10.5	3	idiopathic	36	F, ACE-I, V, D, Hy, Am
10	17	6	4	idiopathic	48	F, ACE-I, D, Am
11	16	8.6	4	idiopathic	48	F, ACE-I, V, D, As
12	11	10	3	ischemic	12	F, ACE-I, V, Wa
Mean	20.7	10.2	3.2	-	34.7	-
SD	6.6	3.1	0.7	-	15.7	-

Blood samples for GH assay were taken every 20 min during the first night of the study (from 10 p.m. of the first day of the study to 6.00 a.m. of the second day of the study). Moreover, blood samples for GH assay were also taken during exogenous GH infusion (at 8.00, 10.00 and 12.00 a.m. of the second day of the study and then every four hours until 8.00 a.m. of the third day of the study). Blood samples for IGF-1 assays were taken at 8.00 a.m. of each of the three days of the study. Standard medical therapy for heart failure was continued unchanged throughout the study. Meals, which were standardised for each patient (8.6 MJ/day; 15% protein, 50% carbohydrates; 35% lipids) were served at 7.00 a.m., 12.00 a.m., 6.00 p.m., of each day of the study. No extra calories were allowed between meals. Chamber dimensions, left ventricular thickness and functional systolic indexes were calculated from M-mode, two-dimensional and doppler echocardiographic (CFM 750 CV;VIGMED, Stockolm, Sweden) measurements with a 3.5 MHz transducer according to the recommendations of the American Society of Echocardiography (9). A symptom-limited upright bicycle (Ergometry System, Siemens Elma, Munich, Germany) exercise test, consisting of a 3 min warming period at 0 watt followed by steps of 10 watts every min has been performed; expired gases have been determined using a mass spectrometer (Medical Gas Analyzer 1100, Marquette, Milwaukee, WI, USA). Exercise test was stopped when patients complained of either severe dyspnea or fatigue. Right heart catheterization was performed in the morning (8.00 a.m.) of the first day of the study after the patients had fasted overnight and had received the usual oral therapy. Pulmonary artery (PAP) and pulmonary capillary (PCWP) pressure: these parameters were measured with a 7.5F Swan-Ganz catheter (Sorenson Research Lab, Salt Lake City, Utah, USA) introduced through the right jugular vein; cardiac output (CO): it was determined in triplicate by the thermodiluition technique with the use of the above mentioned catheter; arterial blood pressure (AP): a brachial artery was cannulated percutaneously with a 20 Gauge (Arrow, Reading, PA, USA) catheter to record arterial blood pressure and to obtain arterial blood samples. All these parameters were measured 30 min after the right heart catheterization (8.00 a.m. of the first day of the study; baseline I), at the end of the control period (8.00 a.m. of the second day of the study; baseline II) and every 4 hours during GH infusion (last measure 8.00 a.m. of the third day of the study; 24-hour GH i.v.).

The same amount of 0.1 UI/Kg body weight/24 hour of biosynthetic recombinant human GH (Humatrope, Eli Lilly, Italy) was administered, through constant continuous intravenuos infusion, to all the patients. This GH dose was chosen based on previous studies showing it to be clearly biologically effective in patients with GH-deficiency (10). The constant GH intravenous infusion was performed with an iv infusion set and pump (Abbott, Lifecare 5000, North Chicago, Illinois, USA) (20 ml/min).

RESULTS

Endocrine data

a) Effects of exogenous GH infusion on GH and IGF-1 levels

Mean GH serum concentration increased, on the average, during infusion five fold as compared to the baseline level (from 2.5±1.4 to 13.6±8.3 μg/L) in our patients; IGF-1 rose of about 50% as compared to the pre-infusion level (from 168.7±61.6 to 248.3± 86.6 μg/L) (Table 2). A certain degree of variability in the effects of exogenous GH on endogenous GH levels and IGF-1 levels was also observed (Figure 1), as previously reported for normal subjects (20). Mean GH levels during the infusion ranged from 3.4 to 33.1 μg/L (from 5 to more than 10 fold increase). Moreover, IGF-1 levels after exogenous GH ranged from 129.9 to 403.9 μg/L (from 15 to 115% increase). Interestingly, the highest IGF-1 levels after the GH infusion were obtained in those patients with higher IGF-1 levels at baseline independently of the serum GH levels obtained during the infusion.

b) Correlations

A negative correlation, however not reaching the statistical significance, has been found between mean nocturnal GH levels and baseline IGF-1 levels ($r=-0.47$, $p=0.124$). Interestingly, negative significant correlations have been found between mean nocturnal GH levels and both post infusion absolute ($r=-0.67$, $p<0.05$) and delta (post- pre-infusion) ($r=-0.58$; $p<0.05$) IGF-1 levels. No significant correlations have been found between several parameters of liver function (albumin, bilirubin and pseudo-cholinesterase) and mean nocturnal GH. However, baseline IGF-1 levels showed a negative significant correlation ($r=-0.76$, $p<0.01$) with total bilirubin and a positive correlation ($r=0.72$, $p<0.01$) with pseudo-cholinesterase.

Table 2. Effects of exogenous GH infusion on GH and IGF-1 levels.

Patient #	Baseline GH (μg/L)	mean GH (μg/L)	GH AUC (μg/L • 24h)	GH peak (μg/L)	Baseline IGF-1 (μg/L)	IGF-1 peak (μg/L)	Δ IGF-1 (peak-baseline)
1	-	-	-	-	181.9	215.2	33.3
2	1.2	9	15325	13.4	101.2	216.8	115.6
3	3.1	14	22704	20.2	97.5	143.2	45.7
4	1.2	7.8	12457	10.2	223.9	373	149.1
5	1.4	18	33082	48.1	274.5	403.9	129.4
6	0.3	7.6	12456	12.4	226.8	348.9	122.1
7	0.1	3.4	5424	4.9	191.2	241.7	50.5
8	3.8	17	25614	35.9	185.9	271.4	85.5
9	0.55	8.5	12690	14.1	169.5	235.1	65.6
10	4.6	20.2	30840	39.9	63.5	129.9	66.4
11	0.7	11	16906	14.3	186	193	7
12	7.6	33.1	53352	53.2	122	207	85
Mean	2.2	13.6	21896	24.2	168.7	248.3	79.6
SD	2.3	8.3	13459	16.9	61.6	86.6	42.9

Hemodynamic data

a) Baseline

The baseline hemodynamic data did show an impaired (11) left ventricular performance in our patients: in fact, mean cardiac index was 2.1±0.8 L/min/m^2. Only four patients had cardiac index ≥ than 2.5 L/min/m^2 (patient # 5, 6, 9 and 12). Mean PAP was 40.3±10.2 mmHg. Moreover, data consisting with pulmonary hypertension (mean PAP > 20 mmHg) were noticed in all but one (#5) of the patients. Systemic and pulmonary vascular resistances were elevated in all the subjects: > 1200 and 250 dine • sec • cm^{-5}, respectively. Importantly, reproducibility of the hemodynamic measurements was confirmed by the observation that all the data at the beginning of the study (8.00 of the first day) were superimposable to those observed at the beginning of the GH infusion (8.00 of the second day of the study) (Table 3).

b) Effects of exogenous GH infusion

Systemic circulation: after 24 hours of continuous GH infusion (Table 3) we found a dramatic (about 50% over the baseline level) increase in mean cardiac index vs baseline levels (3.3±1.2 vs 2.1±0.6 L/min/m^2, $p<0.01$). The cardiac index increase after infusion ranged between 10% and 200% vs baseline. Notably, ten out of the twelve patients had cardiac index ≥ 2.5 L/min/m^2 at the end of the GH infusion; in one other patient cardiac index improved but was not normalized and in one patient only (#12) a slight worsening in cardiac index (from 3 to 2.8 L/min/m^2) was observed during GH infusion. We did not found any differences in both heart rate and systolic arterial pressure, but we observed a significant decrease in diastolic (68.1±7.7

vs 58.7±12.1 mmHg, p<0.05) and mean (81.7±9 vs 74±12.5 mmHg, p<0.05) arterial pressure. Systemic resistances did fall significantly from 1686±518 to 972±292 dyn • sec • cm $^{-5}$ in the whole group of patients (Table 3). Analytical inspection of the data showed that in the large majority of the eleven "responder" patients (ten out of eleven) the cardiac index continued to increase for the whole period of the infusion being the maximal cardiac index value observed at the 24th hour (Figure 1). Consistently, the decrease in both diastolic and mean systemic arterial pressure and in systemic vascular resistances occurred following a similar temporal and qualitative pattern as compared to the increase in cardiac index in each single patient.

Table 3. Effects of GH infusion (24-h GH i.v.) on hemodynamic variables (means ± SD) as compared to two baseline evaluations. Baseline I (8.00 a.m. of the first day of the study). Baseline II (8.00 a.m. of the second day of the study) just before the beginning of the exogenous GH infusion.

Cardiac parameters	Baseline I	Baseline II	p value (Baseline I vs II)	24h GH i.v.	p value (GH vs Baseline II)
Heart Rate (b/min)	70 ±11	70 ±12	n.s.	70.9 ±13.7	n.s.
Cardiac index (L/min/m^2)	2.1 ±0.8	2.1 ±0.6	n.s.	3.3 ±1.2	<0.01
Mean systemic arterial pressure (mmHg)	81.1 ±8.1	81.7 ±9	n.s.	74 ±12.5	<0.05
Systemic vascular resistances (dyn •sec • cm^{-5})	1729 ±521	1686 ±518	n.s.	972 ±292	<0.01
Mean pulmonary arterial pressure (mmHg)	40.3 ±10.2	38.8 ±10.7	n.s.	30.2 ±8	<0.01
Pulmonary capillary wedge pressure (mmHg)	26.5 ±7.3	24.1 ±8.9	n.s.	18.1 ±6	<0.05
Pulmonary vascular resistances (dyn •sec • cm^{-5})	398 ±126	386 ±141	n.s.	172 ±74	<0.01

Pulmonary circulation: after 24 hours of continuous GH infusion an important (about 25% less than baseline parameters) drop in the hemodynamic parameters at the pulmonary circulation level was observed. In fact, a significant decrease in systolic PAP (47.2±14.7 vs 59.7±16.1 mmHg, p<0.01), range 24 to 81 mmHg, diastolic PAP (21.3±6.7 vs 26.9±9.2 mmHg, p<0.05), range 8 to 31 mmHg, and PCWP (18.1±6 vs 24.1±8.9 mmHg, p<0.05), range 6 to 27 mmHg was observed. Mean PAP decrease after infusion ranged between 6.5% and 50% vs baseline. A consensual drop in pulmonary resistances (172±74 vs 386±141 dyn • sec • cm $^{-5}$, range 85 to 271 dyn • sec • cm $^{-5}$) as compared to baseline levels after GH infusion was observed (Table 3).

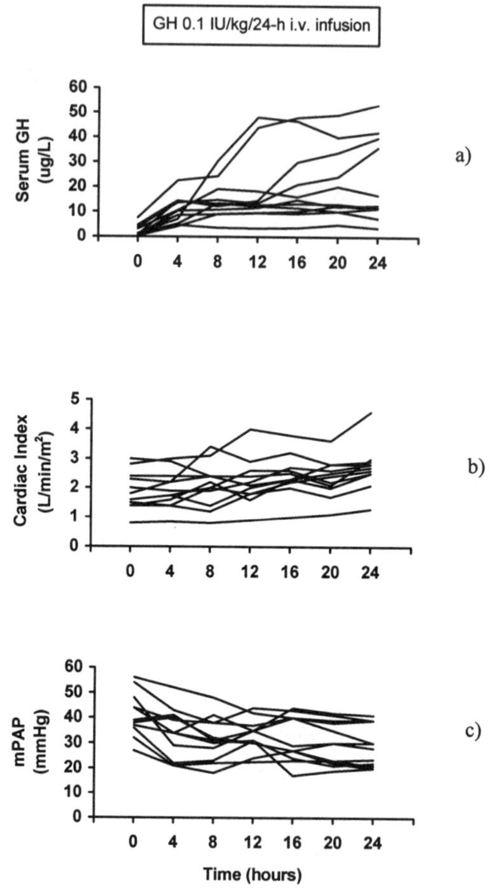

Figure 1. a) Individual serum GH levels during exogenous GH infusion in eleven of the twelve patients (with the exception of patient ≠ 1, see Table 1) of our study; b) Individual cardiac index profile during exogenous GH infusion; c) Individual mean pulmonary arterial pressure (mPAP) profile during exogenous GH infusion.

Interestingly, one of the patients (#9) achieved systolic PAP <30 mmHg and three patients (#5,9 and 11) had mean PAP <20 mmHg after GH infusion. In one patient only (#12) an increase in all the parameters of pulmonary circulation was observed. This patient was the same who did not respond to GH with an increase in cardiac index. Interestingly, most of the patients (n. 8) reached the lowest mean PAP and PCWP between 4 and 8 hours from the beginning of the infusion, maintaining this plateau level for the whole duration of the infusion. In the remaining three "responder" patients mean PAP (Figure 1) and PCWP continued to decrease for the entire period of the infusion being the lowest mean PAP and PCWP value observed at the 24-th hour.

c) Correlations

Baseline IGF-1 levels showed a significant, negative correlation with baseline mean PAP ($r=-0.68$, $p<0.05$) and PCWP ($r=-0.70$, $p<0.05$) and a positive correlation with baseline cardiac index ($r=0.71$, $p<0.05$). Baseline IGF-1 levels showed also a significant negative correlation with absolute, but not delta, mean PAP ($r=-0.63$, $p<0.05$) and mean PCWP ($r=-0.67$, $p<0.05$) after GH infusion.

Post GH infusion IGF-1 levels also negatively correlated with post GH infusion mean PAP ($r=-0.50$, $p=0.09$) and mean PCWP ($r=-0.66$, $p<0.05$). The positive correlation between either baseline or post-infusion IGF-1 and the post-infusion cardiac index ($r=0.40$ and 0.43, respectively) did not reach statistical significance.

Side effects

All the enrolled patients completed the study protocol and no one reported subjective problems during GH infusion. Moreover, no significant cardiovascular or systemic unfavourable effects were observed during the continuous monitoring of vital parameters during GH infusion.

DISCUSSION

Our data (11) show that short-term (24 h) continuous intravenous infusion of recombinant human GH is able to dramatically improve both the left and the right ventricle performance resulting in a dramatic increase in cardiac index together with a reduction in pulmonary hypertension in patients with severe CHF.

That the GH /IGF-1 axis may play a significant role in regulating cardiac function has been suggested by experimental (12) and clinical (6) studies on the cardiovascular effects of GH. In fact, the acute administration of this hormone cause myocardial hypercontractility in normal subjects (6). Moreover, our recent data showed that the acute GH reduction through 24-hour continuous infusion of octreotide (8) and through a single injection of slow-release Lanreotide (13) in acromegalic patients caused significant improvement in work capacity at both anaerobic threshold and during maximal exercise.

The role of the GH/IGF-1 axis in heart failure has been carefully investigated only in the last few years. In fact, spontaneous GH hypersecretion (14) or hyposecretion (15) have been associated to different stages of heart dysfunction in humans. Overall, these data suggested that the pituitary could have a role in the endocrine compensatory response to heart failure. Based on all the above mentioned experimental evidences it was clear that the potential role for GH in the treatment of cardiac failure needed to be carefully tested (5,15).

Therefore, we looked at possible effects of an acute but substained elevation of circulating GH levels, obtained via a 24-h intravenous pump infusion, on cardiovascular function in patients with severe CHF. Our data show that GH was able to acutely increase cardiac index in the large majority of the patients studied; this effect was not only impressive on an average basis (about 50% of improvement vs baseline) but allowed to observe a normalization of this parameter in the large majority of the patients studied. Interestingly, the increase in cardiac output was accompanied by a decrease in pulmonary pressures which was only slightly less impressive than the effect on cardiac index (average decrease of 25%; one fourth of the patients with normalized mean pulmonary pressures). These effects occurred concomitantly with a decrease in systemic and pulmonary resistances. On the basis of these data, it can be hypothesized that GH has very prominent functional effects on the heart even in patients with

CHF. It is likely that GH may increase heart contractility and concomitantly reduce vascular resistances. Interestingly, all the hemodynamic effects of GH started to be observed quite soon (4 to 8 hours after the beginning) during the infusion and were maintained (pulmonary pressures) or continued to improve (cardiac index) throughout the study. The present work confirms the observation, previously made in experimental animals, that GH treatment results in a significant increase in myocardial contractility in vivo (7). Our results are also consistent with previous observations from other investigators that GH can increase myocardial contractility both in vitro (16,17) and in humans (6). Interestingly, unlike other inotropic agents, GH has been shown to increase the contractile performance of myocardium at a lower energy cost (18,19). Further, the present study demonstrates that GH treatment did not affect heart rate in all patients, suggesting that the hemodynamic effect of GH may not be secondary to changes in sympathetic tone. An increase in myocardial contractility associated with low energy consumption and unaltered heart rate could be a favorable effect on a failing heart.

The mechanism by which GH increases myocardial contractility is not entirely clear. Several experimental data suggest that GH may cause at the heart muscle level an increase in the number of active myosin crossbridges, in the amount of calcium available for myofilament activation, in expression of B-myosin heavy chain and in calcium sensitivity of contractile proteins (7). Our data seem to suggest that at least part of these effects may be very rapidly activated by a sustained increase in circulating GH levels. Moreover, our data for the first time demonstrate, what in previous animal studies could only be hypothesized, i.e. that the effects of GH on cardiac performance are independent of the mechanism of cardiac hypertrophy (7). The reduction in afterload is also a beneficial effect of GH in CHF. The present study confirms experimental data showing that treatment with GH produced a significant decrease in systemic vascular resistance in rats with heart failure (7). This finding is also consistent with the clinical observation that GH treatment enhances cardiac output and reduces peripheral vascular resistances in patients with GH deficiency (19). There are two theories regarding the actions of GH: the GH (direct effector) hypothesis and the IGF-1 hypothesis (20). The IGF-1 hypothesis of GH action proposes that GH induces IGF-1 production in the liver and peripheral tissues, and are these IGF-1 that produce many of the effects of GH. In the present study, GH administration resulted in significant elevations in both circulating GH and IGF-1 levels. IGF-1 has been shown to increase the contractility of neonatal rat cardiocytes in vitro (21). In vivo studies have also shown that IGF-1 treatment enhances cardiac output and stroke volume in rats with doxorubicin-induced cardiomyopathy (22). The cardiac effect of GH observed in the present study could be direct and not involve local IGF-1 generation or could be due to IGF-1 being produced either locally in the heart or systemically in the liver. The recent availability of recombinant human IGF-1 could help in clarifying this issue, allowing to study in patients with CHF the acute effect of an increase in circulating IGF-1 indipendent of changes in circulating GH.

Our data give also some interesting insights concerning the possible determinants of the observed positive functional acute cardiovascular effect of GH. In fact, patients characterized by a certain degree of "GH resistance" are likely to have less dramatic endocrine (increase in IGF-1) and hemodynamic responses to GH infusion. Further studies are needed in order to assess if, in the clinical setting, low baseline circulating IGF-1 could represent a good predictor of a weak hemodynamic response to exogenous GH infusion.

We conclude that GH has acute functional effects on the heart in patients with CHF and these effects probably include both an increase in myocardial contractility and a decrease in vascular resistances. Therefore, GH has to be considered, and consequently tested, as a possible new therapeutical option in the short-term treatment of CHF.

REFERENCES

1. Giustina A, Veldhuis JD. Pathophysiology of the neuroregulation of growth hormone secretion in experimental animals and the human. Endocr Rev 1998;19:717-797.
2. Giustina A, Bussi AR, Jacobello C, Wehrenberg WB. Effects of recombinant growth hormone (GH) on bone and intermediary metabolism in patients receiving chronic glucocorticoid treatment with suppressed endogenous GH response to GH-releasing hormone. J Clin Endocrinol Metab 1995;80:122-9.
3. Giustina A, Wehrenberg WB. Growth hormone neuroregulation in diabetes mellitus. Trends Endocrinol Metab 1994;5:73-8.
4. Saccà L, Cittadini A, Fazio S. Growth hormone and the heart. Endocr Rev 1994;15:555-73.
5. Fazio S, Sabatini D, Capaldo B, et al. A preliminary study of growth hormone in the treatment of dilated cardiomyopathy. N Engl J Med 1996;334:809-14.
6. Thuesen L, Christiansen JS, Sorensen KE, Jorgensen JOL, Orslov H, Henningsen P. Increased myocardial contractility following growth hormone administration in normal man. Dan Med Bull 1988;35:193-6.
7. Yang R, Bunting S, Gillet N, Clark R, Jin H. Growth hormone improves cardiac performance in experimental heart failure. Circulation 1995;92:262-7.
8. Giustina A, Boni E, Romanelli G, Grassi V, Giustina G. Cardiopulmonary performance during exercise in acromegaly, and the effects of acute suppression of growth hormone hypersecretion with octreotide. Am J Cardiol 1995;75:1042-7.
9. Sahn DJ, DeMaria A, Kisslo J, Weyman A. Recommendations regarding quantitation in M-mode echocardiography: results of a survery of echocardiographic measurements. Circulation 1978;58:1072-83.
10. Jorgensen JOL, Moller N, Lauritzen T, Christiansen JS. Pulsatile versus continuous intravenous administration og growth hormone (GH) in GH-deficiency patients: effects on circulating insulin-like growth factor-I and metabolic indices. J Clin Endocrinol Metab 1990;70:1616-23.
11. Volterrani M, Desenzani P, Lorusso R, D'Aloia A, Manelli F, Giustina A. Hemodynamic effects of intravenous growth hormone in congestive heart failure. Lancet 1997;349, 1067-1068.
12. Penney DG, Dunbar JC, Baylerian MS. Cardiomegaly and heamodynamics in rats with a transplantable growth hormone-secretin tomuor. Cardiovasc Res 1985;19:270-7.
13. Manelli F, Desenzani P, Boni E, et al. Cardiovascular effects of a single slow release lanreotide injection in patients with acromegaly and left ventricular hypertrophy. Pituitary 1999;2:205-210.
14. Anand IS, Ferrari R, Kalra GS, Wahi PL, Poole-Wilson PA. Edema of cardiac origin: studies of body water and sodium, renal function, hemodynamic indexes and plasma hormones in untreated congestive heart failure. Circulation 1989;80:299-305.
15. Cuneo RC, Wilmhurst P, Lowy C, McGauley G, Sonksen PH. Cardiac failure responding to growth hormone. Lancet 1989;2:838-9.
16. Thorner MO, Vance ML, Horvar E, Kovacs K. The anterior pituitary. In: Wilson JD, Foster DW Editors. Williams Textbook of Endocrinology VIII Edition. Philadelphia, PA: WB Saunders Co;1992:221-310.
17. Timsit J, Mercadier J-J. Effects of chronic growth hormone excess on cardiac contractility and myosine phenotipe in the rat. Acta Peadiatr 1992;383:32-4.

18. Timsit J, Riou B, Bertherat J, et al. Effects of chronic growth hormone hypersecretion on intrinsic contractility, energetics, isomyosin pattern, and myosin adenosine triphosphatase activity of rat left ventricle. J Clin Invest 1990;86:507-15.
19. Mayoux E, Ventura-Clapier R, Timsit J, Behar-Cohen F, Hoffmann C, Mercadier JJ. Mechanical properties of rat cardiac skinned fibers are altered by chronic growth hormone hypersecretion. Circ Res 1993;72:57-64.
20. Caidahl K, Eden S, Bengtsson BA. Cardiovascular and renal effects of growth hormone. Clin Endocrinol (Oxf) 1994;40:393-400.
21. Valcavi R, Gaddi O, Zini M, Lavicoli M, Mellino U, Portioli I. Cardiac performance and mass in adults with hypopituitarism: effects of one year of growth hormone treatment. J Clin Endocrinol Metab 1995;80:659-66.
22. Ambler GR, Johnston BM, Maxwell L, Gavin JB, Gluckmen PD. Improvement of doxorubicin induced cardiomyopaty in rats treated with insulin-like growth factor 1. Cardiovasc Res 1993;27:1368-73.

11
EFFECT OF CHRONIC GROWTH HORMONE TREATMENT IN HEART FAILURE

Maurizio Volterrani[1], Filippo Manelli[2], Andrea Giustina[2]
[1]*Fondazione "S. Maugeri", Cardiology Dept., Gussago Medical Centre, Brescia;*
[2]*Endocrine Section, Dept of Internal Medicine, University of Brescia; Italy*

INTRODUCTION

Congestive heart failure is a multiple aetiology, high prevalence, poor prognosis cardiovascular disorder. Medical treatment of dilated cardiomiophaty is aimed at alleviating the symptoms of heart failure. Growth hormone (GH) and insulin like growth factor (IGF-1) are involved in several physiological processes such as the control of muscle mass and function, body composition and regulation of nutrient metabolism (1,2). The roles of GH and IGF-1 as modulators of myocardial structure and function are well established (3). Receptors for both GH and IGF-1 are expressed by cardiac myocytes; therefore, GH may act directly on the heart or via the induction of local or systemic IGF-1, while IGF-1 may act by endocrine, paracrine or autocrine mechanisms.
Patients with acromegaly have an increased propensity to develop ventricular hypertrophy and cardiovascular diseases as well as an impaired cardiac efficiency is observed in those patients with GH deficiency (4,5).
Animals model of pressure and volume overload have demonstrated up-regulation of cardiac IGF-1 production and expression of GH and IGF-1 receptors, implying that the local regulation of these factors is influenced by hemodynamic changes. Moreover, experimental studies suggest that GH and IGF-1have stimulatory effects on myocardial contractility, possibly mediated by changes in intracellular calcium handling (6).

CLINICAL STUDIES ON GH TREATMENT IN CHF

It has been found that neurohormonal activation is closely related to a poor outcome of CHF (7-10); moreover, drugs that interfere with the neurohormonal changes occurring in CHF (such as ACE- inhibitors and beta-blockers) can positively influence the natural history of the disease (11,12). Like other hormones in CHF, GH axis has also been found to be significantly altered (13). Furthermore, in patients with GH deficiency, left ventricular function is reduced and GH administration normalizes cardiac performance (14). Studies of various models of experimental heart failure suggest that GH may induce myocyte hypertrophy, improve myocardial contractility and change the distribution of myosin isoforms that improve the metabolic efficiency of the heart (15-17). Beneficial effects of GH therapy in patients with chronic heart failure have now been reported in a series of controlled and uncontrolled studies (Table 1) (18). Fazio et al reported that GH administered every second day for 3 months increased septal thickness and left ventricular mass, improved clinical symptoms, exercise capacity and reduced chamber size and end-systolic wall-stress while biochemically inducing a marked increase in IGF-I levels in 7 patients with CHF (19). An improvement of the cardiac index by 57% and a decrease of mean pulmonary artery pressure by 25% was by us shown after a 24h infusion of GH [0.1 IU/Kg] in 12 CHF patients (20). Beneficial clinical responses to GH have also been reported in small groups of CHF patients with increases of body weight, improved functional capacity and exercise tolerance (21,22). However, a study by Frustaci et al. (23) on five patients with severe CHF, showed only mild reduction of left ventricular diameters and minor improvement in left ventricular ejection fraction after 3 months of GH administration (4 IU/day). These initial studies have all been open and uncontrolled. More recently, two placebo controlled studies of GH therapy in CHF patients have been published (24,25). In Isgaard's study, GH was given daily at a dose of 0.25 IU/kg/week to 22 CHF patients. No significant effects on hemodynamic parameters, clinical symptoms, left ventricular mass or exercise tolerance were reported (24). Similarly, a second randomised controlled trial of GH therapy in CHF patients did not show significant benefits on NYHA functional class or on the 6 min walking distance (25).

Table 1. Summary of published studies on GH treatment in patients with CHF.

Author	n° patients	Randomized controlled	Dose	Treatment time
Cuneo [22]	1	No	12IU / day	3 months
Fazio[19]	7	No	4IU 3/week	3 months
Frustaci [23]	5	No	4IU / day	3 months
Volterrani[20]	12	No	0.1IU/kg/24h	24 hours
O'Driscoll[22]	2	No	>10IU / day	1-7 weeks
Osterziel[25]	50	Yes	2IU / day	3 months
Isgaard[24]	22	Yes	2.5 IU / day	3 months

It is not easy to justify the discrepant results of the above mentioned studies of GH administration in CHF patients. It has been previously shown (26) that acquired GH resistance is closely related to CHF, particularly in patients with cardiac cachexia. The different responses to GH therapy might be related to the proportion of patients "resistant" to the effect of the hormone. Growth hormone resistance might be a major determinant of the response to GH treatment, as in this condition its administration would not be effective. Cardiac cachexia might be an important determinant of response to growth hormone therapy, because of its close relationship to GH resistance. In most of the studies previously mentioned (19,20,23-25), it has not been stated whether the patients suffered from body wasting. The only two studies on cachectic patients (21-22) showed significant clinical benefits in this subgroup of patients, but high doses were administered, to possibly override GH resistance. Another theoretical possibility for overcoming GH resistance might be the mode of GH administration as suggested by the same authors. In most studies GH has been given daily, while it has been observed that an intermittent administration is able to produce greater induction of IGF-I mRNA in peripheral tissues (27). Only Fazio et al. (19) used injections every second day, and they were the only who found significant clinical benefits of chronic growth hormone therapy in CHF patients.

CONCLUSIONS

Growth Hormone treatment in CHF patients seems to be promising. In small open studies, acute and chronic GH treatment have demonstrated beneficial effects in patients with heart failure due to ischemic or idiophatic cardiomyopathy. Recently, two randomised, placebo-control studies (24,25), did not show significant GH-mediated improvement in cardiac performance in patients with dilated cardiomyopathy, despite significant increases in IGF-1. Acquired growth hormone resistance, might be an important feature of severe heart failure and explain the different responses to GH therapy seen in different patients (28). Whether GH treatment will finally find a place, and with which modalities, in the treatment of heart failure remains to be established.

REFERENCES

1. Wehrenberg WB, Giustina A. Basic counterpoint: mechanisms and pathways of gonadal steroid modulation of growth hormone secretion. Endocr Rev 1992;13:299-308.
2. Giustina A, Velduis JD. Pathophysiology of the neuroregulation of growth hormone secretion in experimental animals and the human. Endocr Rev 1998;19:717-797.
3. Froesch ER, Schnid C, Scwander J, Zapf J. Action of insulin-like growth factors. Annu Rev Physiol 1985;47:443-467.
4. Manelli F, Desenzani P, Boni E, et al. Cardiovascular effects of a single slow release lanreotide injection in patients with acromegaly and left ventricular hypertrophy. Pituitary 1999;2:205-210.
5. Cuneo RC, Salomon F, McGauley GA, Sonksen PH. The growth hormone deficiency in adults. Clin Endocrinol 1992;37:387-397.
6. Saccà L, Cittadini A, Fazio S. Growth hormone and the heart. Endocr Rev 1994;15:555-573.
7. Swedberg K, Eneroth P, Kjeshus J, Wilhelmsen L, for the CONSENSUS Trial Study Group. Hormones regulating cardiovascular function in patients with severe congestive heart failure and their relation to mortality. Circulation 1990;82:1730-1736.
8. Cohn JN, Levine TB, Olivari MT, et al. Plasma norepinephrine as a guide to prognosis in patients with chronic congestive heart failure. N Engl J Med 1984;311:819-823.
9. Francis GS, Cohn JN, Johnson G, Rector TS, Goldman S, Simon A, for the V-HeFT VA Cooperative Study Group. Plasma norepinephrine, plasma renin activity, and congestive heart failure: relations to survival and the effects of the therapy in V-HeFT II. Circulation 1993;87(suppl VI):VI-40-VI-48.
10. Gottlieb SS, Kukin ML, Ahem D, Packer M. Prognostic importance of atrial natriuretic peptide in patients with chronic heart failure. J Am Coll Cardiol 1989;13:1534-1539.
11. The CONSENSUS Trial Study Group. Effects of enalapril on mortality in severe congestive heart failure: results of the Cooperative North Scandinavian Enalapril Survival Study (CONSENSUS). N Engl J Med 1987;316:1429-1435.
12. Waagstein F, Bristow MR, Swedberg K, et al., for the Metoprolol in Dilated Cardiomyopathy (MDC) Trial Study Group. Beneficial effects of metoprolol in idiopathic dilated cardiomyopathy. Lancet 1993;342:1441-1446.
13. Anker SD, ChuaTP, Ponikowski P, et al. Hormonal changes and catabolic/anabolic imbalance in chronic heart failure and their importance for cardiac cachexia. Circulation.1997;96:526-534.

14. Amato G, Carella C, Fazio S, et al. Body composition, bone metabolism, heart structure and function in growth hormone (GH)-deficient adults before and after GH replacement therapy at low doses. J Clin Endocrinol Metab 1993;77:1671-1676.
15. Yang R, Bunting S, Gillett N, Clark R, Jin H. Growth hormone improves cardiac performance in experimental heart failure. Circulation 1995;92:262-267.
16. Duerr RL, Huang S, Miralizakbar HR, Clark R, Chien KR, Ross J Jr. Insulin-like growth factor-I enhances venticular hypertrophy and function during the onset of experimental heart failure. J Clin Invest 1995;95:619-627.
17. Timsit J, Rion B, Bertherat J, et al. Effects of chronic growth hormone hypersecretion on intrinsic contractility, energetics, isomyosin pattern, and myosin adenosine triphosphatase activity of rat left ventricle. J Clin Invest 1990;86:507-515.
18. Volterrani M, Manelli F, Cicoira MA, Lorusso R, Giustina A. Role of growth hormone in chronic heart failure: therapeutic implications. Drugs 2000, In Press.
19. Fazio S, Sabatini D, Capaldo B, et al. A preliminary study of growth hormone in the treatment of dilated cardiomyopathy. N Engl J Med 1996;334:809-814.
20. Volterrani M, Desenzani P, Lorusso R, d'Aloia A, Manelli F, Giustina A. Hemodynamic effects of intravenous growth hormone in congestive heart failure. Lancet 1997;349, 1067-1068.
21. Cuneo RC, Wilmshurst P, Lowy, McGauley G, Sönksen PH. Cardiac failure responding to growth hormone. Lancet 1989;333:838-839.
22. O'Driscoll JG, Green DJ, Ireland M, Kerr D, Larbalestier RI. Treatment of end-stage cardiac failure with growth hormone. Lancet 1989;I:838-839.
23. Frustaci A, Gentiloni N, Russo MA. Growth hormone in the treatment of dilated cardiomyopathy. N Engl J Med 1996;335:672-673.
24. Isgaard J, Caidahl K, Lomsky M, Hjalmarson Å, Bengtsson BÅ. A placebo controlled study of growth hormone in patients with congestive heart failure. Eur Heart J 1998;19:1704-1711.
25. Osterziel KJ, Strohm O, Schuler J, et al. Randomised, double blind, placebo-controlled trial of human recombinant growth hormone in patients with chronic heart failure due to dilated cardiomyopathy. Lancet 1998;351:1233-1237.
26. Anker SD, Volterrani M, Pflaum C-D, Poole-Wilson PA, Strasburger CJ, Coats AJS. Growth hormone resistance in chronic heart failure. J Am Coll Cardiol 1998(Suppl. A);31:204A.
27. Isgaard J, Carlsson L, Isaksson OGP, Janson JO. Pulsatile intravenous growth hormone (GH) infusion to hypophysectomised rats increases insulin-like growth factor I messenger ribonucleic acid more effectively than continuous growth hormone infusion. Endocrinolgy 1988;123:2605-2610.
28. Giustina A, Volterrani M, Manelli F, et al. Endocrine predictors of the acute hemodynamic effects of growth hormone in congestive heart failure. Am Heart J 1999;137:1035-1043.

12
DOES THE GROWTH FACTOR APPROACH HAVE A FUTURE IN THE TREATMENT OF HUMAN HEART FAILURE?

Luigi Saccà, Antonio Cittadini, Serafino Fazio
Department of Internal Medicine and Cardiovascular Sciences, University Federico II, Naples, Italy

INTRODUCTION

Until recently, the physiologic role of growth hormone (GH) and its tissue effector insulin-like growth factor I (IGF-I) was restricted to the control of linear growth and energetic metabolism, and to the maintenance of skeletal muscle growth and performance (1). Over the past few years, a wealth of data from both animal and human studies have enlarged the spectrum of GH's action even further. GH is now well recognized to be essential for cardiac development and for preserving cardiac structure and performance in adult life (2). IGF-I is rightly included in the list of the many factors that are implicated in cardiac hypertrophy in response to pressure or volume overload.

Interwoven with these effects on the heart are a series of recent observations supporting a role of the GH/IGF-I axis in the development and physiology of the vascular system. Evidence is accumulating that GH and IGF-I are implicated in the control of a variety of vascular functions, including angiogenesis, peripheral vascular resistance, and organ blood perfusion.

As the complex interaction of GH/IGF-I with the cardiovascular system was being unraveled, the idea to use growth factors in the treatment of heart failure crept out (3,4). Despite the promising background, however, the data from clinical studies of GH use in patients with heart failure are quite controversial and do not provide an unequivocal message. For this reason, the present review represents an attempt to re-examine the rationale of the growth factor approach and to discuss the possible reasons why the clinical data are not in line with the expectations generated by the experimental models.

MOLECULAR BASIS FOR GH'S CARDIAC EFFECTS

Before addressing the interaction of GH/IGF-I with cardiac growth and function, it is useful to briefly examine the molecular background that accounts for the cardiac effects of these growth factors. Cardiac tissue expresses the GH receptor very actively, and its stimulation is followed by increased IGF-I expression and content (5). IGF-I is a potent cardiac growth factor. For instance, in neonatal cardiomyocytes, IGF-I stimulates protein synthesis and cell hypertrophy (6). At the same time, IGF-I upregulates some cardiac specific genes, such as beta-myosin heavy chain, myosin light chain-2, and troponin I. This response is entirely mediated by IGF-I (6). In adult cardiomyocytes, IGF-I stimulates myofibril development and sarcomerogenesis (7). Furthermore, in transgenic models of IGF-I overexpression cardiac mass is significantly increased as compared with wild type animals (8).

Binding of GH to its receptors leads to activation of cytoplasmic tyrosine kinases (JAK2 and MAP kinases). Activated JAK2 phosphorylates some members of the STAT family, which translocate into the nucleus, where they trigger the transcription of IGF-I (9)

The signalling pathways activated by the IGF-I-receptor complex are only partially elucidated. They include: the insulin receptor substrate-1 (IRS-1); 2) the Sch proteins; and 3) the phospholipase C pathway. IRS-1, in turn, binds the p85-subunit domain of phosphatidylinositol 3-kinase (10).

Overall, IGF-I appears to be a good candidate as mediator of cardiac growth for the following reasons: 1) there is a specific receptor system in cardiomyocytes; 2) IGF-I upregulates specific genes in cardiomyocytes; 3) IGF-I stimulates *in vitro* protein synthesis in cardiomyocytes, myofibril development, and the accumulation of sarcomere units; and 4) changes in cardiac tissue growth *in vivo* are associated with parallel changes in IGF-I expression.

THE GH/IGF-I STATUS IN HEART FAILURE

Before entering into details of how GH works, it is interesting to examine the GH/IGF-I status in patients with heart failure. The specific questions to ask are: How does the GH/IGF-I axis react to heart failure? What are the GH levels in heart failure? One shoud also get some idea as to the tissue sensitivity to GH and IGF-I in these patients.

If one looks at patients with quite severe, but not end-stage, heart failure, there is evidence for an impaired activity of the GH/IGF-I axis. GH secretion was monitored overnight in patients with class III or IV heart failure (11). The interesting observation was made that in those patients with more severely compromized LV function, the average GH concentration and the peak concentration were considerably lower than in normal subjects. This finding suggested impaired secretion by the pituitary gland. Support for this view comes from another study involving a large number of patients with dilated cardiomyopathy (most in class III) (12). IGF-I was markedly reduced in these patients. In addition, the GH response to growth-hormone-releasing hormone, which is a potent physiologic stimulus, was also reduced. The data demonstrate that in patients with heart failure there is a degree of secretory deficit of the GH/IGF-I axis. These observations are seemingly at variance with a report published in the late 1980's of very high GH levels in patients with advanced heart failure (13). The discrepancy has recently been resolved by an intriguing study conducted at the National Heart and Lung Institute in London (14). In patients with advanced heart failure associated with malnutrition and wasting or frank cardiac cachexia, GH is elevated, but IGF-I is normal or tends to be low. So, the GH:IGF-I ratio is very high. The only explanation for this finding is GH resistance. Noteworthy, GH resistance is the usual response to malnutrition.

The overall message is that, in many patients with heart failure, for one reason or another, GH activity is impaired, either because of defective secretion or because of GH resistance. The data also raise the question as to whether or not the low GH activity contributes to cardiac dysfunction.

It is not easy to answer this question. However, GH plays an important role in cardiac physiology. Patients with congenital GH deficiency have distinct abnormalities in cardiac morphology and function. As already mentioned, the lack of GH impairs cardiac performance (2). There are several case reports of patients with GH deficiency who developed dilated cardiomyopathy and severe heart failure. These patients were totally unresponsive to conventional therapy and only when they were treated with GH, did their clinical status improve dramatically (15-17). In one of them, endomyocardial biopsy was performed before and after GH therapy. The data show that GH treatment increased the cross-sectional area of myocardiocytes and the percent myofibrillar area, indicating that GH induced parallel addition of new sarcomeres (16).

To fully appreciate the impact of GH deficiency on cardiac function, it is necessary to examine data of LV mass and geometry in patients with childhood onset GH deficiency (18). LV mass is remarkably decreased with respect to a matched control group, because of LV wall thinning with no change in cavity dimension. So, the relative wall thickness is significantly reduced. Both LV mass and relative wall thickness are normalized by GH therapy. As predicted by Laplace's law, LV wall stress is remarkably increased in GH deficiency, and partially normalized by GH replacement therapy. Given the high wall stress, it is not surprising that cardiac output is impaired in GH-deficient patients, particularly during physical exercise (19,20).

So, transposed into a larger scale, the geometric alterations of the LV in GH deficiency are similar to those seen in heart failure and the consequences are equally similar and they include the high wall stress and ventricular dysfunction. In addition, the abnormalities caused by GH deficiency are reversed by specific treatment, which demonstrates the continuing ability of GH to modulate cardiac tissue growth and the plasticity of cardiac tissue in adult life. This was the basic consideration that led to the growth factor approach to heart failure.

GH AND SKELETAL MUSCLE

There is extensive evidence regarding the involvement of skeletal muscle in heart failure. The abnormalities encompass muscle ultrastructure, biochemistry, energetics, and consequently performance. The mechanism(s) underlying these abnormalities is not well known, although likely candidates are chronic muscle underperfusion, deconditioning, prevalence of hormonal catabolic forces, and more recently activation of cytokines, in particular TNF-α.

The effects of GH in patients with heart failure have never adequately addressed. It is not unreasonable to hypothesize that GH may be favorable for several considerations. First, GH allows for the maintenance of normal skeletal muscle growth and performance. Patients with GH deficiency often complain of fatigue. They have a sedentary lifestyle and a low exercise capacity. Second, GH is able to reverse each of the abnormalities associated with heart failure when they occur in other disease states, particularly in GH deficiency: 1) GH stimulates protein synthesis and this effect coupled with the anabolic effect of IGF-I may initiate reversal of muscle atrophy (21); 2) heart failure entails changes in muscle composition with a shift from type I to type II fibers. GH has been shown to be able to raise the percentage of type I fibers (22); 3) the FFA mobilizing effect of GH leads to increased availability of this substrate to the muscle where GH raises the oxidation rate and energy output (23); 4) both GH and IGF-I increase muscle perfusion through NO-mediated mechanisms (24) and this may counteract the chronic underperfusion associated with heart failure.

GH AND CARDIAC GROWTH IN HEART FAILURE

From a growth perspective, the very basic event in heart failure is the growth response triggered by diastolic overloading secondary to the loss of contractile mass and power. Because the prevalent mechanical stimulus is diastolic overstretching of cardiomyocytes, the growth response occurs essentially by virtue of series addition of new sarcomeres. This inevitably entails fiber elongation and ventricular dilatation. And because ventricular dilatation is not accompanied by proportional wall thickening, for reasons inherent to Laplace law, wall stress increases further, the diastolic growth stimulus is maintained, and a vitious cycle is perpetuated.

Current approaches in the treatment of heart failure are based on measures that act through a variety of mechanisms, among which possibly reduction of wall stress is pre-eminent. Wall stress reduction is achieved by unloading the failing heart either through an effect on ventricular pressures (nitrates, ACE inhibitors, beta-blockers) or through surgical remodeling of the left ventricle.

The alternative approach using growth factors aims at inducing additional hypertrophy in an attempt to raise wall thickness and to make wall stress decline. So, the real questions are whether GH is indeed able to initiate a growth response in the failing heart and whether this response bears physiologic features in terms of ventricular geometry and structure.

Evidence from experimental studies show that GH or IGF-I or the combination of the two increases LV mass (3,25). Similarly, data from human volunteers treated for 4 weeks with GH demonstrate an increase in LV mass and relative wall thickness (26). Perhaps more interesting is the model of uncomplicated acromegaly, in which long-term exposure of the heart to elevated GH concentration causes a net increase in the relative wall thickness with attendant fall in wall stress (27). These data, however, do not necessarily imply that a similar response must occur in heart failure.

The effects of GH/IGF-I in heart failure have been assessed in the post-infarction rat model. The data show an effect of these growth factors to raise LV wall thickness and to improve ventricular dysfunction (2,28,29). It is possible that GH and IGF-I also act through growth-independent, purely functional mechanisms. For instance, GH and IGF-I improve myocardial contractility and relaxation, and reduce peripheral vascular resistance (30-33).

With regard to the effects of GH in human heart failure, the data available are quite inconsistent. A variety of small, uncontrolled studies have shown a beneficial effect both in acute and chronic clinical settings (34-40). There are also reports showing no effect at all (41,42). There is only one large, placebo-controlled study that included 50 patients with idiopathic dilated cardiomyopathy (43). This study demonstrated that GH raises LV mass and wall thickness. There was, however, no apparent functional or clinical benefit.

THE CARDIAC GROWTH RESPONSE TO GH

Two questions are of particular relevance: 1) Is there any real advantage in increasing cardiac mass in heart failure? and 2) How physiologic is cardiac growth response to GH?

Regarding the first question, it depends on the geometric features that the growth response confers to the left ventricle. If an increase in wall thickness occurs without cavity enlargement, the Laplacian burden is attenuated, and this contributes to unload the failing heart. In the majority of studies using experimental models of heart failure, GH induced a geometrically favorable pattern of cardiac growth. There is no conclusive evidence that GH does the same in human heart failure.

Regarding the second question, the following considerations suggest that the growth response to GH is physiologic: 1) Cardiac tissue growth occurs with maintenance of proportionality between cardiomyocytes and capillary bed. This aspect has been investigated in the postinfarction model and in the rapid ventricular pacing model of heart failure. Despite the induction of cardiac hypertrophy, capillary density was either unaffected or increased by GH (28,44). 2) In short-term studies in animal models of heart failure, GH or IGF-I did not increase myocardial interstitial tissue (25,28). In a more recent study examining the effect of more prolonged (three months) GH administration, the collagen concentration was consistently decreased both in the right and in the left ventricle (45). Accordingly, GH was reported to decrease the amount of collagen I and fibronectin in the noninfarcted areas of postinfarcted left ventricles (46). This effect was associated with improvement of systolic and diastolic function. Particularly relevant is the observation that in patients with acromegaly diastolic function is well preserved, despite cardiac hypertrophy, if the disease duration does not exceed five years (27). 3) In several models of heart failure, GH induced the expression of SR-Ca^{2+} ATPase (47,48), increased the availability of Ca^{2+} to the myofilaments (49) and preserved the activity of the SR ryanodine receptors (50). These effects at molecular level were mirrored by improved ventricular function. 4) GH and IGF-I do not upregulate the gene program usually associated with heart failure (51). In isolated cardiomyocytes, IGF-I induced expression of myosin light chain-2 and troponin I (6). Interestingly, the state of phosphorylation of these proteins was enhanced in a transgenic model of IGF-I overexpression (52).

REASONS FOR THE INCONSISTENT RESULTS IN HEART FAILURE

It is not easy to understand why GH and IGF-I have proven beneficial in experimental heart failure and in small, uncontrolled human studies, whereas the data from larger or controlled studies have been less impressive or disappointing. Trying to explain the difference among clinical studies only by the experimental design (placebo-controlled and double-blind *vs.* open) is too simplistic an explanation. This may be a contributory factor but it cannot be the only reason. There are several, well-performed animal studies based on unquestionable designs that show clear-cut, positive effects of GH or IGF-I on the failing heart, both in terms of LV geometry and function. In addition, GH has been shown to influence favorably the myocardial capillary density, the collagen:cardiomyocytes growth ratio, and some molecular pathways, e.g. SR-Ca^{2+} ATPase. Obviously, these findings are unlikely to be the mere consequence of a placebo effect.

It is possible that the discrepancy between experimental and human studies is related to the dose of GH used. In animal studies, GH is given in much larger doses than in clinical studies. It should be also considered that in many patients with heart failure, the activity of the GH/IGF-I axis is impaired and IGF-I concentration is low. When GH is administered to these patients, IGF-I concentration is simply normalized or stabilizes in the high physiologic range. Another potential reason for the different results may be the dose schedule. When the most impressive response to GH was described, the hormone had been administered every second day. Given the pulsatile nature of GH secretion, one cannot exclude that the temporal pattern of administration and tissue exposure may be relevant.

Finally, patients with heart failure may undergo GH resistance. This mostly occurs in end-stage heart failure or whenever there are concomitant nutritional problems. The GH resistance syndrome is characterized by high GH levels and inappropriately low IGF-I levels, due to the reduced tissue ability to generate IGF-I. So, in some patients/studies the dose of GH used might have been too low to overcome GH resistance. Interestingly, in a recent report of the acute effect of GH in heart failure, it was observed that the lower the IGF-I concentration at baseline the weaker the hemodynamic response to GH (52). This may be explained by a state of GH resistance, as reflected by the low IGF-I level, and the blunted responsiveness of the system to GH's action. The interpretation is supported by the finding that those patients with the lowest IGF-I levels at baseline responded to GH administration with the least increase in IGF-I concentration.

REFERENCES

1. Saccà L, Cittadini A, Fazio S. Growth hormone and the heart. Endocr Rev 1994;15:555-573.
2. Saccà L. Growth hormone. A newcomer in cardiovascular medicine. Cardiovasc Res 1997;36:3-9.
3. Duerr RL, Huang S, Miraliakbar HR, Chien K, Ross J Jr. Insulin-like growth factor 1 enhances ventricular hypertrophy and function during the onset of experimental cardiac failure. J Clin Invest 1995;95:619-627.
4. Saccà L, Fazio S. Cardiac performance: growth hormone enters the race. Nature Med 1996;2:29-31.
5. Mathews LS, Enberg B, Norstedt G. Regulation of rat growth hormone receptor gene expression. J Biol Chem 1989;17:9905-9910.
6. Ito H, Hiroe M, Hirata Y, et al. Insulin like growth factor-I induces hypertrophy with enhanced expression of muscle specific genes in cultured rat cardiomyocytes. Circulation 1993;87:1715-1721.
7. Donath MY, Zapf J, Eppenberger-Eberhardt M, Froesch EF, Eppenberger HM. Insulin-like growth factor 1 stimulates myofibril development and decreases smooth muscle alpha-actin of adult cardiomyocytes. Proc Natl Acad Sci 1994;91:1686-1690.
8. Li Q, Li B, Wang X, et al. Overexpression of insulin-like growth factor-1 in mice protects from myocyte death after infarction, attenuating ventricular dilatation, wall stress, and cardiac hypertrophy. J Clin Invest 1997;100:1991-1999.
9. Darnell JE, Kerr IM, Stark GR. JAK-STAT pathways and transcriptional activation in response to INFs and other extracellular signalling proteins. Science 1994;264:1415-1421.
10. Foncea R, Andersson M, Ketterman A, et al. Insulin-like growth factor-I rapidly activates multiple signal transduction pathways in cultured rat cardiac myocytes. J Biol Chem 1997;272:19115-19124.
11. Giustina A, Lorusso R, Borghetti V, Bugari G, Misitano V, Alfieri O. Impaired spontaneous growth hormone secretion in severe dilated cardiomyopathy. Am Heart J 1996;131:629-632.
12. Broglio F, Gianotti L, Fubini A, et al. Activity of GH/IGF-I axis in patients with dilated cardiomyopathy. Ann Endocrinol 1997;58(suppl 1):S88.
13. Anand IS, Ferrari R, Kalra GS, Wahi PL, Poole-Wilson PA, Harris PC. Edema of cardiac origin: studies of body water and sodium, renal function, hemodynamic indexes and plasma hormones in untreated congestive heart failure. Circulation 1989;80:299-305.

14. Anker SD, Chua TP, Ponikowski P, et al. Hormonal changes and catabolic/anabolic imbalance in chronic heart failure and their importance for cardiac cachexia. Circulation 1997;96:526-534.
15. Cuneo RC, Wilmshurst P, Lowy C, McGauley G, Sønksen PH. Cardiac failure responding to growth hormone. Lancet. 1989;1:838-9.
16. Frustaci A, Perrone GA, Gentiloni N, Russo MA. Reversible dilated cardiomyopathy due to growth hormone deficiency. Am J Clin Pathol. 1992;97:503-11.
17. Fazio S, Biondi B, Sabatini D, et al. Long-term growth hormone deficiency as a cause of cardiomyopathy and its reversibility with specific replacement therapy. J Clin Endocrinol Metab. 1996;81:887-890.
18. Fazio S, Cittadini A, Sabatini D, et al. Growth hormone and heart performance: a novel mechanism of cardiac wall stress regulation in humans. Eur Heart J 1997;18;340-347.
19. Cittadini A, Cuocolo A, Merola B, et al. Impaired cardiac performance in GH-deficient adults and its improvement after GH replacement. Am J Physiol 1994;267:E219-E225.
20. Cuocolo A, Nicolai E, Colao A, et al. Improved left ventricular function after growth hormone replacement in patients with hypopituitarism: assessment with radionuclide angiography. Eur J Nucl Med 1996;23:390-394.
21. Fryburg DA. Insulin-like growth factor-I exerts growth hormone- and insulin-like actions on human muscle protein. Am J Physiol 1994;267:E331-E336.
22. Nagulesparen M, Trickey R, Davies MJ, Jenkins JS. Muscle changes in acromegaly. Lancet 1976;II:914-915.
23. Leung K-C, Ho KKY. Stimulation of mitochondrial fatty acid oxidation by growth hormone in human bibroblasts. J Clin Endocrinol Metab 1997;82:4208-4213.
24. Fryburg DA. N^G-monomethyl-arginine inhibits the blood flow but not the insulin-like response of forearm muscle to IGF-I. J Clin Invest 1996;97;1319-1328.
25. Cittadini A, Strömer H, Katz SE, et al. Differential cardiac effects of growth hormone and insulin-like growth factor-1 in the rat. A combined in vivo and in vitro evaluation. Circulation 1996;93:800-809.
26. Cittadini A, Ehrnborg C, Longobardi S, et al. Growth hormone activates cardiac growth in normal subjects: a multicenter, double-blind, placebo-controlled study. 81st Annual Meeting, The Endocrine Society, San Diego 1999. Abstract OR4-2.
27. Fazio S, Cittadini A, Biondi B, et al. Cardiovascular effects of short-term acromegaly: Evaluation by Doppler-echocardiography in a strictly selected group of young patients with uncomplicated disease. J Clin Endocrinol Metab 2000;85:179-182.
28. Duerr RL, McKirnan D, Gim RD, Clark RG, Chien KR, Ross JJr. Cardiovascular effects of insulin-like growth factor-1 and growth hormone in chronic left ventricular failure in the rat. Circulation 1996;93:2188-2196.
29. Cittadini A, Grossman JD, Napoli R, et al. Growth hormone attenuates early left ventricular remodeling and improves cardiac function in rats with large myocardial infarction. J Am Coll Cardiol 1997;29:1109-1116.
30. Vetter U, Kupferschmid C, Lang D, Pentz S. Insulin-like growth factors and insulin increase the contractility of neonatal rat cardiocytes in vitro. Bas Res Cardiol 1988;83:647-654.
31. Yang R, Bunting S, Gillett N, Clark R, Jin H. Growth hormone improves cardiac performance in experimental heart failure. Circulation 1995;92:262-267.
32. Strömer H, Cittadini A, Douglas PS, Morgan JP. Exogenously administered growth hormone and IGF-I alter intracellular calcium handling and enhance cardiac

performance: in vitro evaluation in the isolated isovolumic buffer perfused rat heart. Circ Res 1996;79:227-236.

33. Cittadini A, Ishiguro Y, Strömer H, et al. Insulin-like growth factor-1 but not growth hormone augments mammalian myocardial contractility by sensitizing the myofilament to Ca^{2+} through a wortmannin-sensitive pathway. Studies in rat and ferret isolated muscles. Circ Res 1998;83:50-59.
34. Fazio S, Sabatini D, Capaldo B, et al. A preliminary study of growth hormone in the treatment of dilated cardiomyopathy. N Engl J Med. 1996;334:809-814.
35. O'Driscoll JG, Green DJ, Ireland M, Kerr D, Larbalestier RI. Treatment of end-stage heart failure with growth hormone. Lancet 1997;349:1068.
36. Volterrani M, Desenzani P, Lorusso R, D'Aloia A, Manelli F, Giustina A. Hemodynamic efects of intravenous growth hormone in coongestive heart failure. Lancet 1997;349:1067-1068.
37. Beer N, Tortoledo F, Beer R, Pinedo M. Beneficial effects of growth hormone in patients with congestive heart failure. II Meeting of the Working Group on Heart Failure, Cologne, abs P294.
38. Genth-Zotz S, Zotz RJ, Geil S, Voigtlander T, Meyer J, Darius H. Recombinant growth hormone therapy in patients with ischemic cardiomyopathy. Effects on hemodynamics, left ventricular function and cardiopulmonary exercise capacity. Circulation 1999;99:18-21.
39. Spallarossa P, Rossettin P, Minuto F, et al. Evaluation of growth hormone administration in patients with chronic heart failure secondary to coronary artery disease. Am J Cardiol 1999;84:430-433.
40. Donath MY, Sutsch G, Yan XW, et al. Acute cardiovascular effects of insulin-like growth factor I in patients with chronic heart failure. J Clin Endocrinol Metab 1998;83:3177-3183.
41. Khardori R, Tabaren A, Taylor GJ, Falcone M. Acute effects of growth hormone infusion in patients with advanced congestive heart failure. Lancet 1997;349:1842.
42. Isgaard J, Bergh C-H, Caidahl K, Lomsky M, Hjalmarson A, Bengtsson BA. A placebo-controlled study of growth hormone ion patients with congestive heart failure. Eur Heart J 1998;19:1704-1711.
43. Osterziel KJ, Strohm O, Schuler J, et al. Randomised, double-blind, placebo-controlled trial of human recombinant growth hormone in patients with chronic heart failure due to dilated cardiomyopathy. Lancet 1998;351:1233-1237.
44. Spinale FG, Houck WV, Pan LC. Growth hormone supplementation with developing congestive heart failure: Effects on left ventricular cellular structure and composition. Circulation 1998;98(Suppl I):180.
45. Bruel A, Oxlund H. The effect of long-term growth hormone administration on the collagen concentration, relative amounts of type I and III collagen and pyridinoline collagen cross-links in rat hearts. Growth Hormone & IGF Research 1998;8:328.
46. Grimm D, Cameron D, Griese DP, Riegger GAJ, Kromer EP. Differential effects of growth hormone on cardiomyocyte and extracellular matrix protein remodeling following experimental myocardial infarction. Cardiovasc Res 1998;40:297-306.
47. Tajima M, Weinberg EO, Bartunek J, et al. Treatment with growth hormone enhances contractile reserve and intracellular calcium transients in myocytes from rats with postinfarction heart falure. Circulation 1999;99:127-134.

48. Iannini JP, Houck WV, Pan LC, et al. Growth hormone supplementation with developing congestive heart failure: effects on left ventricular myocyte contractile processes. Circulation 1998;98:I-3.
49. Kinugawa S, Tsutsui H, Ide T, et al. Positive inotropic effect of insulin-like growth factor-I on normal and failing cardiac myocytes. Cardiovasc Res 1999;43:157-164.
50. Ueyama T, Ohkusa T, Yano M, Matsuzaki M. Growth hormone preserves cardiac sarcoplasmic reticulum Ca^{2+} release channels (ryanodine receptors) and enhances cardiac function in cardiomyopathic hamsters. Cardiovasc Res 1998;40:64-73.
51. Tanaka N, Ryoke T, Hongo M, et al. Effects of growth hormone and IGF-I on cardiac hypertrophy and gene expression in mice. Am J Physiol 1998;275:H393-H399.
52. Redaelli G, Malhotra A, Li B, et al. Effects of constitutive overexpression of insulin-like growth factor-1 on the mechanical characteristics and molecular properties of ventricular myocytes. Circ Res 1998;82:594-603.
53. Giustina A, Volterrani M, Manelli F, et al. Endocrine predictors of acute hemodynamic effects of growth hormone in congestive heart failure. Am Heart J 1999;137:1035-1043.

13

EXERCISE AND GROWTH HORMONE SECRETION

Johannes D. Veldhuis and Arthur Weltman
University of Virginia Health Sciences Center - General Clinical Research Center
Charlottesville, Virginia

INTRODUCTION

Exercise favors enhanced cardiovascular conditioning, increased muscle strength and maintenance of bone mineral accretion. Many facets of these physiological adaptations are dependent in part on adequate trophic hormones, such as testosterone, growth hormone (GH) and insulin-like growth factor type I (IGF-I) (1-6). In addition, acute exercise stimulates the release of GH as well as that of multiple other hormones and metabolites (7-11). Sustained physical training can elevate circulating concentrations of GH as well as IGF-I and GH-binding protein (12). Importantly, a number of clinical variables affect the magnitude and nature of the GH response to exercise (Table I). Neuroendocrine mechanisms underlying exercise-facilitated GH secretion are complex, and likely include somatostatin withdrawal, GHRH release and possibly co-secretagogue actions (1). Aging and obesity characteristically impair basal GH secretion and the GH response to an exercise stimulus (1,11,13-17). However, in the healthy young adult with normal body composition, vigorous exercise is a potent and effective GH secretagogue (9,18-20). Indeed, exercise is nearly unique in its ability to generate recurrent GH elevations serially throughout the same day (21), and to show escape from GH autonegative feedback (22). In healthy endurance-trained athletes, exercise of short duration (10 min) at intensities exceeding the lactate threshold typically evoke substantial GH release (23), whereas in less well-trained individuals an exercise-intensity dose-response relationship is evident even below the lactate threshold when a 30 min exercise bout is used (18,24). The present review highlights some of these regulatory aspects of the GH-IGF-I axis under acute or chronic exercise drive.

Table I. Clinical variables modulating exercise-stimulated secretion of GH.

- Acute: threshold and/or intensity dependence of the exercise stimulus
- Chronic: intensity, type and/or duration of training
- Fitness level
- Gender
- Age
- Obesity/body composition
- Drugs/medications

PRIMARY ISSUES

Acute exercise

Exercise constitutes one of the more remarkable physiological stimuli to GH release (1,6,9,11,18-21,23,25-28). Some minimum intensity and/or duration of exercise appears to be required to evoke GH secretion in any individual. How consistent this minimum is within subjects is not thoroughly demonstrated. However, exercise stimulus requirements vary significantly among different individuals (26). Typically GH release is delayed for 10-15 minutes following the onset of a 30 min interval of exercise (7,9,19,21,25,26), and attains a maximum toward the end of exertion (25,26) or slightly thereafter (11) (Figure 1). We recently reported that independently of exercise intensity the GH peak occurs approximately 20 min after cessation of a 30-min exercise bout (29). In one study, ten or more minutes of high-intensity exercise were required to elicit GH release consistently (23). There is also significant nonuniformity in the magnitude of the serum GH concentration peak triggered by exercise in healthy individuals (6,11,23,25-27). This secretory variability is evident in response to diverse nonexercise stimuli as well, possibly reflecting short-term fluctuations of endogenous somatostatin inhibitory tone (1). Other reported differences in the amount of GH secreted during exercise may reflect unequal gender representation and variable ages and body composition among participants (1,17).

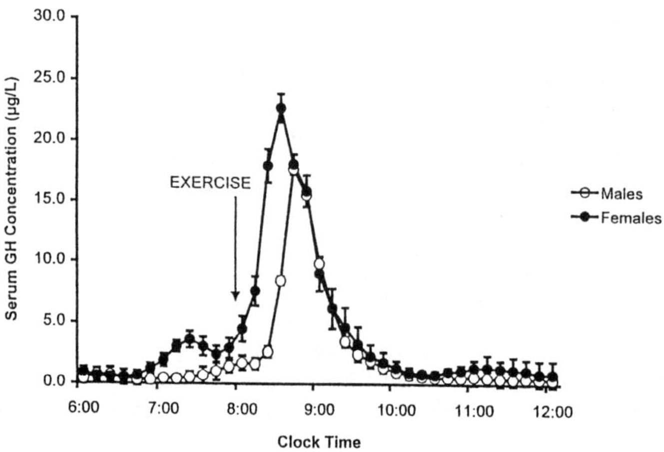

Figure 1. Serial serum GH concentrations in a group of 9-10 young healthy men and women sampled at 10-min intervals for 2 hours before, during 30 min of submaximal exercise (50% of the increment between the individual lactate threshold and the volunteer's maximal oxygen consumption rate) and for 3.5 hours after exercise. Data are the mean ± SEM at each time point, Adapted with permission from (18).

The amount of GH released in response to exercise is intensity dependent (9,11,18,19,23,24,26,30,31). Chawalbinska *et al.* (1996) evaluated the impact of successively incremented exercise intensities within a single study session in 12 young male endurance-trained athletes (32). This analysis identified each subject's lactate threshold as the particular intensity level associated with abruptly increased GH release. However, lactate infusion itself does not evoke GH secretion (33), thus indicating that some other physiological correlate of lactate accumulation is the relevant stimulus. Results in two other studies also suggested that a threshold-like intensity level must be exceeded before exercise induces GH release (34). Viru *et al.* (35) estimated an activation threshold of 50% of the maximal oxygen consumption rate. However, another investigation in less well-trained young men subjected to randomly ordered exercise sessions of independent intensities carried out on separate days demonstrated a strong linear relationship between GH release and exercise intensity across a full range of effort levels both below and above the lactate threshold (18) (Figure 2). A linear relationship between exercise intensity and GH release has also been observed in young women. Preliminary analyses suggest that the mean slope and intercept are greater in young women than young men (36). Few if any comparable data have been obtained to date under this randomized experimental design in young women or older adults (30).

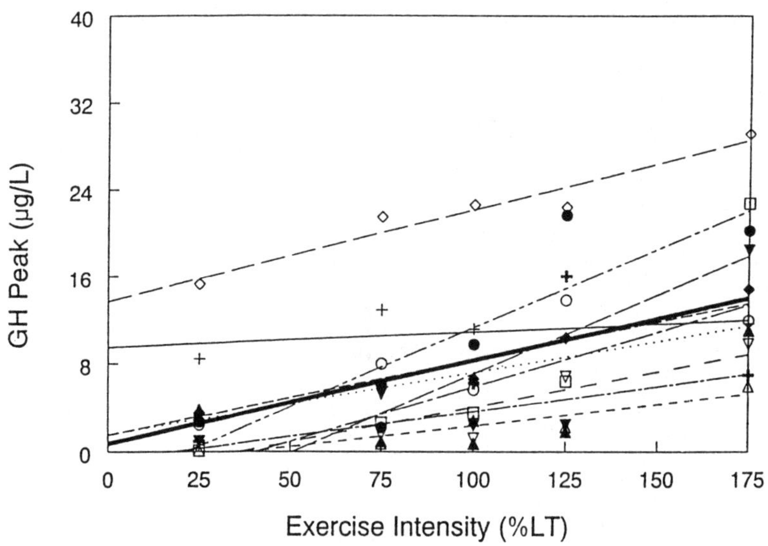

Figure 2. Linear relationship between individualized exercise intensity (x-axis) and peak serum GH concentration (y-axis) in a cohort of healthy young men (mean age 26 years, N = 10). The dark solid line is the group mean regression. Exercise intensity is normalized as a percentage of each volunteer's individual lactate threshold (30).

New analytical techniques have aided recently in efforts to quantitate underlying GH secretion and half-life simultaneously from the exercise-stimulated serum GH concentration profile (13,21,37-40). Application of these analytical tools has revealed that underlying GH secretion rates are truly elevated in response to exercise. Two reports have estimated that if anything the half-life of GH is unchanged or declines slightly during exercise compared with values reported at rest; e.g. in one study the GH half-life did not change (18) and in another slightly decreased from ~19 min (rest) to ~16 min (exercise) (27). A shorter [or unchanged (18)] GH half-life in the face of rising plasma GH concentrations clearly denotes accentuated GH secretion (41), assuming that the plasma volume (or GH distribution space) does not fall proportionately (42). In one clinical assessment of the ability of exhaustive compared with moderate-intensity exercise to enhance GH release, the latter was found (unexpectedly) to be more effective (8). However, in this experiment, the high-intensity study session was always preceded by moderate and low intensity exercise bouts earlier in the same day (8). This fixed sequence may be confounding, since in another investigation there was progressive attenuation of exercise-induced GH secretion after several daily exposures to a schedule of repeated exercise stress without intervening rest intervals of at least 2-3 hours (17). On the other hand, two other more recent studies have demonstrated that as many as three daily 30-minute exercise episodes separated by 1-4 hours do not impede, but actually progressively augment, GH release (21,23). One of these studies may have been confounded in experimental design, since the duration of successive exercise stimuli was not randomly ordered, but rather allowed to increase unidirectionally (23). In an alternative experimental design comprising three randomly ordered exercise sessions each of the same intensity and duration, recurrent episodes of exercise of 30 min separated by one versus 3.5 hours were equivalent in evoking GH secretion in young men (21). In fact, successive constant-load exercise stimuli tended to progressively heighten GH release. Notably, the exercise stimulus is distinctive in this regard, possibly because it can override in some measure GH autonegative feedback (22). Some external factors, such as elevated environmental temperature, do not greatly influence exercise-stimulated GH release; e.g. Brennen *et al.* observed that GH secretion is elicited equivalently by exercise at 40° and 23°C (43). On the other hand, in one small study, GH secretion did not occur when exercise was carried out in the cold (44). This observation allows the conjecture that a core (internal) thermogenic effect of exercise contributes in some manner to stimulation of GH release (1).

Impact of physical training on exercise-stimulated GH release

Hartley *et al.* (8) noted that seven weeks of physical training attenuated exercise-induced GH release after graded-intensity work. Analogously, Weltman *et al.* (20) observed a decline in the GH response to constant load exercise after 3 weeks of intensive training. This was related to constant load exercise occurring at a lower relative intensity after training. In contrast, in a study in swimmers, Bonifaz et al. reported a unidirectional trend for GH release to increase as the training season progressed (45). The mode of exercise, increases in exercise intensity as the season progressed, and/or unequal (thermal) cooling effects associated with swimming might explain these differences. In a longer-term study, 1 year of intensive running training in young women doubled 24-hour integrated serum GH concentrations and GH pulse amplitude even on a non-exercising day, but only in women who trained at exercise intensities exceeding their individual lactate thresholds (6) (Figure 3). The sustained rise in circulating GH concentrations even on non-exercise days may be relevant to long-term training benefits by physiologically elevating tissue exposure to GH. In another study, two weeks of intensive physical training elevated fasting serum concentrations of GH-binding protein by a mean increment of 100 pM and IGF-I by 90 μg/L (12). Analogously, in a 3-month study of obese postmenopausal women, administration of recombinant GH in combination with moderate exercise facilitated loss of intraabdominal (visceral) fat over the effect of exercise alone (46). Thus, GH (of either endogenous or exogenous origin) might enhance the beneficial effects of exercise. However, far more study is required to assess this conjecture adequately.

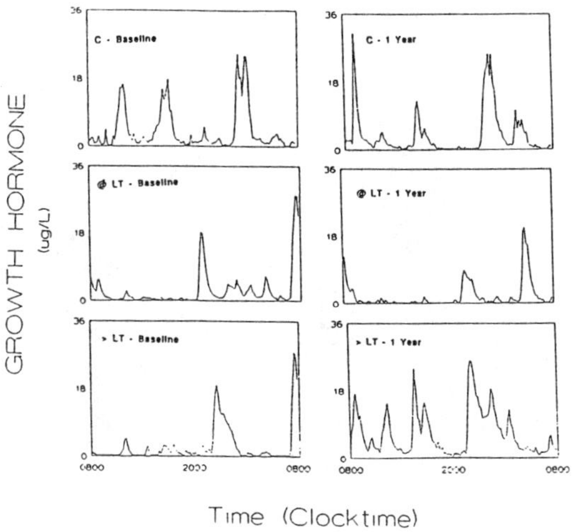

Figure 3. Twenty-four hour serum GH concentration profiles collected at a 20-min sampling interval in a control (untrained, top) and two endurance-trained women (middle and bottom) before (left) and after (right) one year of exercise training. Training was carried out either below (middle) or at/above (bottom) the individual lactate threshold. Adapted with permission from (6).

Whereas exercise is a potent physiological stimulus of GH release, the neuroendocrine mechanism(s) that mediate this effect are not fully understood. Several major considerations are given in Table II. As reviewed recently (1), multiple hypothalamic neurotransmitters, such as dopamine, norepinephrine, acetylcholine, serotonin, GABA and opioids, participate in the control of basal and possibly also exercise-induced GH release (Figure 4). Several of these control pathways are reviewed below in relation to their putative contributions to the exercise stimulus. In fact, no single neuroregulatory mechanism is known to activate GH secretion during exercise. Accordingly, the present perspective is that several CNS mechanisms converge on the final common pathway(s) that drive and suppress GH release, namely, GHRH and somatostatin.

Table II. Possible CNS mechanisms of exercise-stimulated GH release.

• Somatostatin withdrawal	
• GHRH secretion	
• GHRP-receptor activation (putative)	
• Other secretagogues	Alpha 2-adrenergic (?)
	Galanin
	GABA
	Neuropeptide Y
	Opiates
	Other peptides
• Escape from GH and/or IGF-I autonegative feedback.	

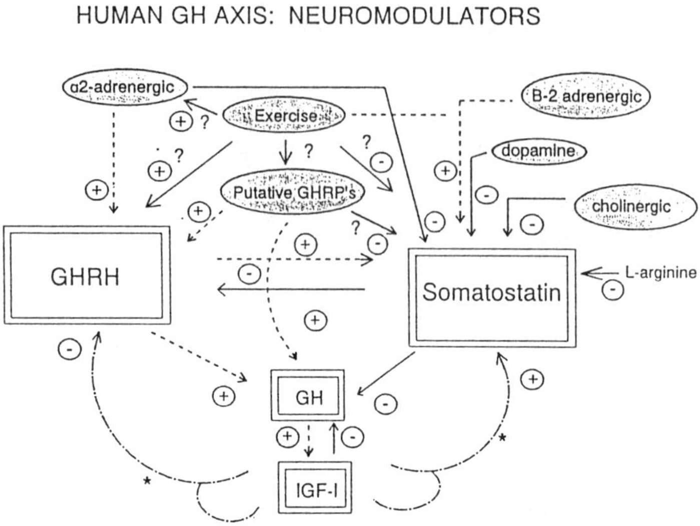

Figure 4. Schematized overview of CNS neurotransmitter control of the GH-IGF-I axis, including autofeedback by GH and IGF-I (asterisks). The putative impact of the exercise stimulus within this control network is shown.

Adrenergic pathways

Both alpha and beta-adrenergic inputs prominently regulate GH secretion (1,47,48). Each can also modulate exercise-induced GH release. For example, investigations in the human using phentolamine (a nonselective alpha receptor-antagonist), metoprolol (a cardioselective beta-2 receptor inhibitor) and propranolol (a nonselective beta-1 and beta-2 receptor antagonist) have shown that beta-2 receptor blockade consistently enhances exercise-stimulated GH release (47-50). In contrast, alpha-receptor inhibitors influenced GH secretion variably during exercise. For example, in two independent studies under ostensibly similar experimental conditions, phentolamine either substantially suppressed (49), or minimally affected (47), exercise-induced GH secretion. At rest, alpha-2 adrenergic receptors are believe to mediate stimulatory pathways driving GH release (1).

Cholinergic

The cholinergic (muscarinic) system is a significant positive modulator of GH secretion, acting presumptively via withdrawal of somatostatin. The indirect cholinergic agonist, pyridostigmine, administered during exercise additively increased GH release (51). Additivity of GH responses would suggest independent regulatory pathways. Since the other dominant input to GH control is GHRH stimulation, one could infer that exercise promotes endogenous GHRH secretion and/or action, but has a lesser effect on somatostatin withdrawal. This facilitative effect of pyridostigmine on exercise-stimulated GH release was corroborated by Thompson *et al.* (27). These authors further noted that the opiate-receptor antagonist, naltrexone, did not impair the stimulatory effect of cholinergic predominance. In view of the foregoing, one might infer that cholinergic innervation promotes somatostatin withdrawal, whereas exercise stimulates alpha 2-adrenergic drive, withdraws beta-2 adrenergic inhibition, and/or modulates other (nonopiatergic) neurons that evoke heightened GHRH release. Consistent with (but not proof of) this interpretation is the ability of phentolamine (an alpha-receptor antagonist) to impair exercise-stimulated GH release in one of two studies (above). Conversely, atropine (a muscarinic antagonist, considered to release somatostatin) blocked exercise-evoked GH secretion (52). Nonetheless, the effects of atropine are not so readily interpreted, since heightened somatostatin release would be expected to restrain GH secretion in response to essentially all stimuli, including GHRH. In addition, the notion that pyridostigmine impedes hypothalamic somatostatin secretion is inferred from studies in the rodent, whereas, in sheep, the indirect cholinergic agonist, neostigmine, stimulates hypothalamic GHRH secretion into portal blood (1).

Opiates

The role of the endogenous opiate system in exercise-stimulated GH secretion is not clear. For example, in one analysis, a high dose of naloxone, an opiate receptor antagonist, abolished GH released otherwise provoked by 20 minutes of high-intensity exercise in competitively trained athletes (53). On the other hand, in another assessment of untrained human volunteers, naloxone had no effect on the exercise stimulus (54). This disparity might arise if endurance training itself favored emergence of an inhibitory opioid pathway that restrained GH release during exercise. In untrained volunteers, concurrent administration of valproic acid (a GABAergic agonist) and naloxone eliminated GH release otherwise elicited by exercise (54). One can speculate from these data that a GABAergic mechanism might interact with that of endogenous opioids to modulate GH release during exercise. The validity of this postulate will need to be confirmed or refuted by further study. However, Thompson et al. (27) reported that naltrexone did not alter resting serum GH concentrations, exercise-enhanced GH release, or the rise in GH concentrations stimulated by pyridostigmine. This finding speaks against a major interaction between opiaterigc and cholinergic pathways in transducing the effect of exercise.

GHRH and GHRP's

GHRP's comprise novel synthetic, non-GHRH, non-somatostatin oligopeptide GH secretagogues or their nonpeptidyl mimetics; e.g. Hexarelin, GH-releasing peptide-2 (GHRP-2), GHRP-6, MK0677 etc. (55-58). These potent GH agonists typically synergize with GHRH (1,59-61). Both the hypothalamus and pituitary gland express receptors for GHRP-like ligands, although no endogenous GHRP molecules has yet been identified. Although the exact in vivo mechanism of action of GHRP's is not clarified (1,58,62), we recently observed that exercise augments the GH-releasing action of this class of secretagogues (9,63). Exercise also heightened GH release during infusion of L-arginine in both men and women(63). L-arginine is a putative somatostatin inhibitor (64). Exercise further facilitated GH secretion in men stimulated simultaneously with infusions of both GHRP-2 and L-arginine, even these two agents together already show synergy (9) (Figure 5). Such findings point to novel or multiple interacting mechanisms of exercise-induced GH secretion (1).

IGF-I and its binding proteins

Data do not definitively characterize changes in IGF-I and its binding protein in acute compared with sustained exercise in both sexes. One careful prospective study showed a rise in plasma total IGF-I in 10 young men following two weeks of endurance training (12). The concentration of high-affinity GH binding protein in plasma increased concomitantly in this experiment.

Influence of Gender

The dependencies of daily GH secretion on percentage body fat, age and physical fitness are two-fold stronger in young men than comparably aged premenopausal women (65,66). For example, 24-hour integrated serum GH concentrations are more negatively influenced by increased percentage body fat or age, and more positively affected by physical fitness, in young men than comparably aged women (66). This disparity disappears in older age (67). Whether this remarkable gender difference is attributable to higher estrogen concentrations in women, or higher testosterone levels in men, is not known, although estrogen and aromatizable androgen are dominant stimuli of GH secretion (68). Gender-based differential control of GH secretion in response to an exercise stimulus has not been investigated thoroughly in the human. However, in several acute exercise studies, young men and women responded similarly (9,25,30,63). Resting GH release is typically higher in women than men (65,69,70). The GH area under the curve attained during a single acute exercise bout is also greater in women (19). In addition, our recent observations suggest that the slope and intercept of the exercise intensity (5 different levels) GH relationship is higher in young women compared with young men (36). Another consistent gender difference is women's tendency to reach their zenith GH concentration earlier (at 24 vs 32 min) than men in response to an exercise stimulus (30). At rest (nonexercising), women also exhibit a greater GH response to L-arginine infusion than men (63,64), suggesting relatively higher somatostatin restraint in the female basally (9). Since unstimulated serum GH concentrations are higher in women (1), we conjecture that GHRH release is also relatively elevated in women compared with men. An alternative hypothesis that the GH distribution volume is smaller and/or the GH half-life longer in women was refuted recently (42). In addition, we found no evident gender distinction in peak GH release during bolus GHRP-2 (1 μg/kg i.v.) infusion (9,63), akin to earlier observations using a (near-maximal) dose of this peptidyl agonist (58).

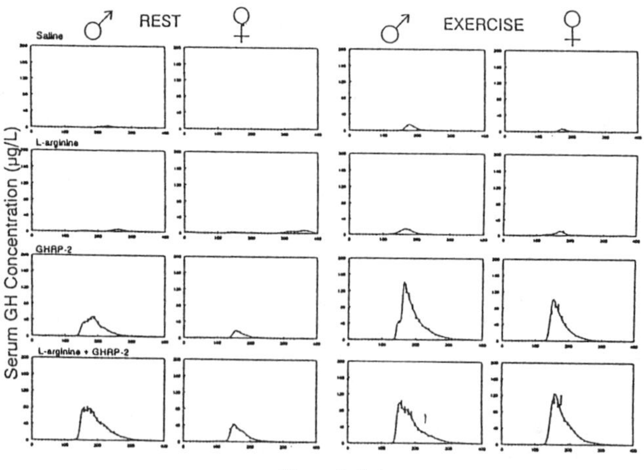

Figure 5. Impact of infusion of L-arginine (to withdraw endogenous somatostatin, at 0830 h) and/or GHRP-2 (a novel GH -releasing peptide, at 0900 h) on resting and exercise-stimulated GH release sampled every 10 min in one healthy young man and women. The 30-min exercise stimulus (Figure 1) was initiated at 120 min with or without secretagogue infusions administered concomitantly. Adapted with permission from (9).

Overall neuroendocrine regulation of exercise-stimulated GH release

Available clinical studies, albeit incomplete, point to possible modulatory roles of beta-adrenergic, alpha-adrenergic, cholinergic (muscarinic), opiatergic and GABA-ergic neurotransmitter pathways in supervising exercise-stimulated GH secretion (1,27,48,50,52-54). In aggregating these diverse data, we schematize in Figure 4 (above) several putative neurotransmitter pathways and their known interactions that may govern exercise-induced GH release. In addition, in recent mechanistics studies, we have observed that the joint effect of exercise and L-arginine is less than that of combined exercise and GHRP-2 in young men (9) (Figure 5). This observation could indicate that exogenous somatostatin withdrawal is not the sole hypothalamic mechanism capable of mediating the exercise stimulus, since L-arginine but not GHRP's appear to act by limiting release of this dominant inhibitor (1). The foregoing findings thus also allow for (unproven) roles of other putative co-secretagogues in mediating GH release driven by exercise; e.g. galanin, neuropeptide Y, a presumptive GHRP-like ligand, etc. (1). The nature of such accessory neuromodulators is not known, nor are their mechanisms of regulation.

CONCLUSIONS

Exercise appears to stimulate GH secretion via novel and/or multifactorial mechanisms in men and women. GH release seems to be exercise-intensity dependent in young men and women provided 30 min of acute exercise is performed. For shorter bouts of exercise (10-min), GH release appears to be intensity-dependent, so long as the lactate threshold is exceeded. Age and increased adiposity suppress, whereas pyridostigmine, L-arginine and GHRP facilitate, GH release evoked by the exercise stimulus. Repeated exercise in the course of a single day will continue to elicit GH release. Exercise combined with recombinant GH injections might facilitate visceral fat loss in postmenopausal women (46) and GH deficient adults (71,72), whereas GH supplementation without exercise may increase muscle strength in GH deficient adults (73) GH supplementation does not seem to enhance the increase in muscle mass or force conferred by resistance training alone in older men (74) and women (75). A one-year prospective study in young women showed that running exercise at an intensity exceeding the individual lactate threshold approximately doubles 24-hr serum GH concentrations even on non-exercise days (6,66,76). The mechanisms underlying such long-term adaptive changes in the GH axis are not known, but do not likely involve a fall in plasma IGF-I concentrations, since IGF-I levels may rise with training (12). Table III highlights several additional unresolved queries in this arena, which ongoing clinical studies will likely begin to clarify.

Table III. Selected unresolved issues regarding exercise and GH secretion.

- How do the mechanism(s) of exercise-driven GH release differ in acute versus chronic exercise (e.g. role of IGF-I, central adrenergic outflow, opiatergic tone, muscle proprioceptive feedback, etc.)?
- What is the magnitude of GH's role in exercise-stimulated bone-mineral accretion, skeletal muscle hypertrophy and cardiovascular conditioning?
- What mechanisms subserve the putative synergy among exercise, sex-steroids, GH and IGF-I in promoting tissue anabolism?

For example, how neuroendocrine mechanisms differ in acute versus chronic stimulation of GH release by exercise is not well articulated. The precise physiological roles of the rise in GH and IGF-I with exercise are also not established. In this light, data do not yet delineate the extent of target-tissue interactions among GH, IGF-I and sex steroids (77) in the overall anabolism promoted by endurance training. Finally, the full nature of and the mechanisms subserving gender differences in exercise-driven GH release also remain incompletely defined (65).

ACKNOWLEDGMENTS
We thank Patsy Craig for her skillful preparation of the manuscript, Paula Azimi and Dr. Laurie Wideman for some of the scientific illustrations, and Dr. Cyril Bowers for insights offered in his reading of the GHRP section.

Supported in part by NIH RO1 AG14799 and General Clinical Research Center (GCRC) grant RR-00847 from NCRR (Bethesda, MD).

REFERENCES

1. Giustina A, Veldhuis JD. Pathophysiology of the neuroregulation of GH secretion in experimental animals and the human. Endocr Rev 1998;19:717-797.
2. Mauras N, Rogol AD, Haymond MW, Veldhuis JD. Sex steroids, growth hormone, IGF-I: neuroendocrine and metabolic regulation in puberty. Hormone Res 1996;45:74-80.
3. Fryburg DA, Weltman A, Jahn LA, et al. Androgenic modulation of the growth hormone-IGF axis and its impact on metabolic outcomes. Presented at the Serono Symposia USA, Inc. entitled "Sex-Steroid Interactions with Growth Hormone" held in Naples, Florida, October 22-25. Norwell, Massachusetts: Serono Symposia USA; 1999.
4. Juul A, Bang P, Hertel NT, et al. Serum insulin-like growth factor-I in 1030 healthy children, adolescents, and adults: relation to age, sex, stage of puberty, testicular size, and body mass index. J Clin Endocrinol Metab 1994;78:744-752.
5. Silva ME, Magalhaes RA, Gonfinetti NV, et al. Effects of testosterone on growth hormone secretion and somatomedin-C generation in prepubertal growth hormone deficient male patients. Brazilian J Med Biol Res 1992;25:1117-1126.
6. Weltman A, Weltman JY, Schurrer R, Evans WS, Veldhuis JD, Rogol AD. Endurance training amplifies the pulsatile release of growth hormone: effects of training intensity. J Appl Physiol 1992;76(6):2188-2196.
7. Weltman A, Weltman JY, Kanaley JA, Rogol AD, Veldhuis JD, Hartman ML. Repeated bouts of exercise alter the blood lactate-ratings of perceived exertion relationship. Med Sci Sports Exerc 1998;30:1113-1117.
8. Hartley LH, Mason JW, Hogan RP, et al. Multiple hormonal response to graded exercise in relation to physical training. J Appl Physiol 1972;33(5):602-606.
9. Weltman A, Wideman L, Weltman JY, et al. Gender impact on the GH response to exercise. Presented at the Serono Symposia USA, Inc. entitled "Sex-Steroid Interactions with Growth Hormone" held in Naples Florida, October 22-25. Norwell, Massachusetts: Serono Symposia USA; 1999.
10. Kanaley JA, Boileau RA, Bahr JA, Misner JE, Nelson RA. Substrate oxidation and GH responses to exercise are independent of menstrual phase and status. Med Sci Sports Exerc 1992;24:873-880.
11. Sutton J, Lazarus L. Growth hormone in exercise: comparison of physiological and pharmacological stimuli. J Appl Physiol 1976;41(4):523-527.

12. Roelen CA, de Vries WR, Koppeschaar HP, Vervoorn C, Thijssen JH, Blankenstein MA. Plasma insulin-like growth factor-1 and high affinity growth hormone-binding protein levels increase after two weeks of strenuous physical training. Int J Sports Med 1997;18:238-241.
13. Veldhuis JD, Liem AY, South S, et al. Differential impact of age, sex-steroid hormones, and obesity on basal versus pulsatile growth hormone secretion in men as assessed in an ultrasensitive chemiluminescence assay. J Clin Endocrinol Metab 1995;80:3209-3222.
14. Iranmanesh A, Lizarralde G, Veldhuis JD. Age and relative adiposity are specific negative determinants of the frequency and amplitude of growth hormone (GH) secretory bursts and the half-life of endogenous GH in healthy men. J Clin Endocrinol Metab 1991;73:1081-1088.
15. Veldhuis JD, Iranmanesh A, Weltman A. Elements in the pathophysiology of diminished growth hormone (GH) secretion in aging humans. Endocrine 1997;7:41-48.
16. Martin FC, Yeo AL, Sonksen PH. Growth hormone secretion in the elderly: aging and the somatopause. Bailliere's Clin Endocrinol Metab 1997;11:223-250.
17. Parkin JM. Exercise as a test of growth hormone secretion. Acta Endocrinol (Copenh.) 1986;279:47-50.
18. Pritzlaff CJ, Wideman L, Weltman JY, et al. Impact of acute exercise intensity on pulsatile growth hormone (GH) release in men. J Appl Physiol 1999;87:498-504.
19. Wideman L, Weltman JY, Shah N, Story S, Veldhuis JD, Weltman A. The effects of gender on exercise-induced growth hormone (GH) release. J Appl Physiol 1999;87:1154-1162.
20. Weltman A, Weltman JY, Womack CJ, et al. Exercise training decreases the growth hormone (GH) response to acute constant-load exercise. Med Sci Sport Exerc 1997;29:669-676.
21. Kanaley JA, Weltman JY, Veldhuis JD, Rogol AD, Hartman ML, Weltman A. Human growth hormone response to repeated bouts of aerobic exercise. J Appl Physiol 1997;83:1756-1761.
22. Wideman L, Weltman JY, Weltman A, Patterson M, Mistry DJ, Veldhuis JD. Effect of growth hormone (GH) infusion on GH secretion during rest and exercise. Presented at the 80th Endocrine Society Annual Meeting, New Orleans, Louisiana, June 24-27, 1998.
23. Felsing NE, Brasel JA, Cooper DM. Effect of low and high-intensity exercise on circulating growth hormone in men. J Clin Endocrinol Metab 1992;75:157-162.
24. Vanhelder WP, Goode RC, Radomski MW. Effect of anaerobic and aerobic exercise of equal duration and work expenditure on plasma growth hormone levels. Eur J Appl Physiol 1984;52:255-257.
25. Lassarre C, Girard F, Durand J, Raynaud J. Kinetics of human growth hormone during submaximal exercise. J Appl Physiol 1974;37:826-830.
26. Raynaud J, Capderou A, Martineaud JP, Bordachar J, Durand J. Intersubject variability in growth hormone time course during different types of work. J Appl Physiol 1983;55(6):1682-1687.
27. Thompson DL, Weltman JY, Rogol AD, Metzger D, Veldhuis JD, Weltman A. Cholinergic and opioid involvement in release of growth hormone during exercise and recovery. J Appl Physiol 1993;75:870-878.
28. Weltman A, Seip RL, Snead D, et al. Exercise training at and above the lactate threshold in previously untrained women. Int J Sports Med 1992;13(3):257-263.
29. Weltman A, Pritzloff CJ, Wideman L, Abbott RD, Hartman ML, Veldhuis JD. Growth hormone (GH) release may be mediated by heightened central adrenergic activity.

Presented at the 81st Annual Endocrine Society Meeting, San Diego, California, June 12-15, 1999.
30. Pritzlaff CJ, Wideman L, Weltman JY, et al. Effects of exercise intensity on growth hormone (GH) release. To Presented at the 45th Annual Meeting of the American College of Sports Medicine, Orlando, Florida, 1998.
31. Vigas M, Nemeth S, Jurcovicova JM, Komadel L. The importance of lactate in exercise-induced growth hormone release in man. Horm Res 1992;166-169.
32. Chwalbinska-Moneta J, Krysztofiak F, Ziemba A, Nazar K, Kaciuba-Uscilko H. Threshold increases in plasma growth hormone in relation to plasma catecholamine and blood lactate concentrations during progressive exercise in endurance-trained athletes. Eur J Appl Physiol Occup Physiol 1996;73:117-120.
33. Luger A, Watschinger B, Deuster P, Svoboda T, Clodi M, Chrousos GP. Plasma growth hormone and prolactin responses to graded levels of acute exercise and to a lactate infusion. Neuroendocrinology 1992;56:112-117.
34. Chang FE, Dodds WG, Sullivan M, Kim MH, Malarkey WB. The acute effects of exercise on prolactin and growth hormone secretion: comparison between sedentary women and women runners with normal and abnormal menstrual cycles. J Clin Endocrinol Metab 1986;62:551-556.
35. Viru A. Pancreatic hormones and the somatotropin-somatomedian system. Hormones in Muscular Activity. Boca Raton: CRC Press 1985:61-75.
36. Weltman A, Pritzlaff CJ, Wideman L, Abbott RD, Hartman ML, Veldhuis JD. Impact of acute exercise intensity on growth hormone (GH) release in women. Presented at the 46th Annual Meeting of the American College of Sports Medicine, Seattle, Washington, 1999.
37. Veldhuis JD, Carlson ML, Johnson ML. The pituitary gland secretes in bursts: appraising the nature of glandular secretory impulses by simultaneous multiple-parameter deconvolution of plasma hormone concentrations. Proc Natl Acad Sci USA 1987;84:7686-7690.
38. Veldhuis JD, Johnson ML. Deconvolution analysis of hormone data. Methods Enzymol 1992;210:539-575.
39. Veldhuis JD. How does one get at glandular secretion, when only hormone concentrations are measured? Clin Endocrinol 1997;46:397-400.
40. Bright GM, Veldhuis JD, Iranmanesh A, Baumann G, Maheshwari H, Lima J. Appraisal of growth hormone (GH) secretion: evaluation of a compositive pharmacokinetic model that discriminates multiple components of GH input. J Clin Endocrinol Metab 1999;84:3301-3308.
41. Veldhuis JD, Lassiter AB, Johnson ML. Operating behavior of dual or multiple endocrine pulse generators. Am J Physiol 1990;259:E351-E361.
42. Shah N, Aloi J, Evans WS, Veldhuis JD. Time-mode of growth hormone (GH) entry into the bloodstream and steady-state plasma GH concentrations, rather than sex, estradiol, or menstrual-cycle stage, primarily determine the GH elimination rate in healthy young women and men. J Clin Endocrinol Metab 1999;84:2862-2869.
43. Brenner IK, Zamecnik J, Shek PN, Shephard RJ. The impact of heat exposure and repeated exercise on circulating stress hormones. Eur J Appl Physiol Occup Physiol 1997;76:445-454.
44. Weeke J, Gundersen HJG. The effect of heating and central cooling on serum TSH, GH, and norepinephrine in resting normal man. Acta Physiol Scand 1983;117:33-39.

45. Bonifazi M, Bela E, Lupo C, Martelli G, Zhu B, Carli G. Influence of training on the response to exercise of adrenocorticotropin and growth hormone plasma concentrations in human swimmers. Eur J Appl Physiol Occup Physiol 1998;78:394-397.
46. Thompson JL, Butterfield GE, Gylfadottir UK, et al. Effects of human growth hormone, insulin-like growth factor I, and diet and exercise on body composition of obese postmenopausal women. J Clin Endocrinol Metab 1998;83:1477-1484.
47. Sutton J, Lazarus L. Effect of adrenergic blocking agents on growth hormone responses to physical exercise. Horm Metabol Res 1974;6:428-429.
48. Uusitupa M, Siitonen O, Harkonen M, et al. Modification of the metabolic and hormonal response to physical exercise by beta-blocking agents. Ann Clin Res 1982;14:165-167.
49. Hansen AP. The effect of adrenergic blockade on the exercise-induced serum growth hormone rise in normals and juvenile diabetics. J Clin Endocrinol Metab 1971;33:807-812.
50. Maclaren NK, Taylor GE, Raiti S. Propanolol-augmented, exercise-induced human growth hormone release. Pediatrics 1975;56:804-807.
51. Cappa M, Grossi A, Benedetti S, Drago F, Loche S, Ghigo E. Effect of the enhancement of the cholinergic tone by pyridostigmine on the exercise-induced growth hormone release in man. J Endocrinol Invest 1993;16:421-424.
52. Casanueva FF, Villaneuva L, Cabranes JA, et al. Cholinergic mediation of growth hormone secretion elicited by arginine, clonidine and physical exercise in man. J Clin Endocrinol Metab 1984;59:526-530.
53. Moretti C, Fabbri A, Gnessi L, et al. Naloxone inhibits exercise-induced release of PRL and GH in athletes. Clin Endocrinol 1983;18:135-138.
54. Coiro V, Volpi R, Maffei ML, et al. Opioid modulation of the gamma-aminobutyric acid-controlled inhibition of exercise-stimulated growth hormone and prolactin secretion in normal man. Eur J Endocrinol 1984;131:50-55.
55. Bowers CY. GH releasing peptides: structure and kinetics. J Pediatr Endocrinol Metab 1993;6:21-31.
56. Bowers CY. A new dimension on the induced release of growth hormone in obese subjects. J Clin Endocrinol Metab 1993;76:817-818.
57. Svensson J, Lonn L, Jansson J-O, et al. Two-month treatment of obese subjects with the oral growth hormone (GH) secretagogue MK-677 increases GH secretion, fat-free mass, and energy expenditure. J Clin Endocrinol Metab 1998;83:362-369.
58. Bowers CY. Synergistic release of growth hormone by GHRP and GHRH: scope and implication. In: Bercu BB, Walker RF, eds. Growth Hormone Secretagogues in Cilnical Practice. New York: Marcel Dekker, Inc.; 1998:1-25.
59. Bowers CY, Reynolds GA, Durham D, Barrera CM, Pezzoli SS, Thorner MO. Growth hormone (GH)-releasing peptide stimulates GH release in normal men and acts synergistically with GH-releasing hormone. J Clin Endocrinol Metab 990;70:975-982.
60. Tiulpakov AN, Brook CG, Pringle PJ, Peterkova VA, Volevodz NN, Bowers CY. GH responses to intravenous bolus infusions of GH releasing hormone and GH releasing peptide 2 separately and in combination in adult volunteers. Clin Endocrinol 1995;43:347-350.
61. Casanueva FF, Micic D, Pombo M, et al. Role of the new growth hormone-releasing secretagogues in the diagnosis of some hypothalamopituitary pathologies. Metab Clin Exp 1996;45:123-126.

62. Ghigo E, Arvat E, Bellone J, Ramunni J, Camanni F. Neurotransmitter control of growth hormone secretion in humans. J Pediatr Endocrinol Metab 1994;6:263-266.
63. Wideman L, Weltman JY, Shah N, et al. The effects of arginine and GHRP-2 on resting and exercise-induced growth hormone (GH) release. Presented at the 79th Endocrine Society Annual Meeting, Minneapolis, MN, June 11-14,1997.
64. Merimee TJ, Rabinowitz D, Fineberg SE. Arginine-initiated release of human growth hormone. Factors modifying the response in normal men. N Engl J Med 1969;281:1434-
65. Veldhuis JD. Gender differences in secretory activity of the human somatotropic (growth hormone) axis. Eur J Endocrinol 1996;134:287-295.
66. Weltman A, Weltman JY, Hartman ML, et al. Relationship between age, percentage body fat, fitness and 24 hour growth hormone release in healthy young adults: effects of gender. J Clin Endocrinol Metab 1994;78:543-548.
67. Ho KKY, Evans WS, Blizzard RM, et al. Effects of sex and age on the 24-hour profile of growth hormone secretion in man: importance of endogenous estradiol concentrations. J Clin Endocrinol Metab 1987;64:51-58.
68. Iranmanesh A, Veldhuis JD. Clinical pathophysiology of the somatotropic (GH) axis in adults. In: Veldhuis JD, eds. Endocrinology and Metabolism Clinics of North America. Philadelphia, PA: W.B. Saunders; 1992:783-816.
69. Winer LM, Shaw MA, Baumann G. Basal plasma growth hormone levels in man: new evidence for rhythmicity of growth hormone secretion. J Clin Endocrinol Metab 1990;70:1678-1686.
70. Van den Berg G, Veldhuis JD, Frolich M, Roelfsema F. An amplitude-specific divergence in the pulsatile mode of GH secretion underlies the gender difference in mean GH concentrations in men and premenopausal women. J Clin Endocrinol Metab 1996;81:2460-2466.
71. Korytko AI, Zeitler P, Cutler L. Developmental regulation of pituitary growth hormone-releasing hormone receptor gene expression in the rat. Endocrinology 1996;137:1326-1331.
72. Bengtsson B, Staffan E, Lonn L. Treatment of adults with growth hormone deficiency and recombinant human GH. J Clin Endocrinol Metab 1993;76:309-317.
73. Johannsson G, Grimby G, Sunnerhagen KS, Bengtsson B-A. Two years of growth hormone (GH) treatment increase isometric and isokinetic muscle strength in GH-deficient adults. J Clin Endocrinol Metab 1997;82:2877-2884.
74. Zachwieja JJ, Yarasheski KE. Does growth hormone therapy in conjunction with resistance exercise increase muscle force production and muscle mass in men and women aged 60 years or older? Physiol Ther 1999;79:76-82.
75. Taafee DR, Jin IH, Vu TH, Hoffman AR, Marcus R. Lack of effect of recombinant human growth hormne (GH) on muscle morphology and GH-insulin-like growth factor expression in resistance-trained elderly men. J Clin Endocrinol Metab 1996;81:421-425.
76. Rogol AD, Weltman A, Weltman JY, et al. Durability of the reproductive axis in eumenorrheic women during one year of endurance training. J Appl Physiol 1992;72(4):1571-1580.
77. Giustina A, Cappelli C. Interaction between testosterone and growth hormone replacement therapy. In: Lamberts SWJ, eds. The Diagnosis and Treatment of Pituitary Insufficiency. Bristol: BioScientific, Ltd.; 1997:181-208.

14
GROWTH HORMONE AND SKELETAL MUSCLE FUNCTION

Ettore C. Degli Uberti, Paola Franceschetti, Maria R. Ambrosio
Department of Biomedical Sciences and Advanced Therapies (Section of Endocrinology), University of Ferrara, Ferrara, Italy

INTRODUCTION

It is well established that skeletal muscle is a target tissue for GH, and there is a wealth of literature concerning the influence played by GH on the musculoskeletal system. It is now widely recognized that GH-Insulin-like growth factor-I (IGF-I) axis has an important role in controlling the growth and differentation of skeletal muscle. Recently, this topic was extensively reviewed by Florini et al (1). As early as 1934, it was reported that administration of anterior pituitary extracts in rats increases muscle mass (2). These observations were confirmed by Plattener and Reed (3), who found that, in rats, treatment with "anterior pituitary growth factor" leads to an increase in muscle mass, but does not affect muscle performance. Accordingly, rats treated with 500 μg GH/day for 3 weeks gained significantly more than controls, and their quadriceps muscles were 15-40% heavier, although muscle performance did not change (4). In hypophysectomized animals GH administration has induced an increase in muscle mass and activity of polyribosomes, indicative of enhanced muscle protein synthesis (5). The strong link between GH and skeletal muscle growth has been convincingly demonstrated by Palmer et al (6) who provided clear evidence that intact rats treated with a polyclonal antiserum to rat GH exhibit a marked reduction in weight, total protein and RNA content of hind limb muscle. Futhermore, it is also well known that hypophysectomized rats have a decrease in the levels of mRNA for all forms of myosin heavy chain (MCH) that is partially remedied by GH (7). The mode of action on skeletal muscle, however, still remains unclear. Although, the somatomedin hypothesis implies that many of the biological activities of GH are mediated by IGFs, which could regulate the proliferation and differentiation of skeletal muscle cell, a direct effect of GH on muscle cannot be excluded. Indication of a direct effect of GH has been provided by Nixon and Green (8) using cultured mouse myoblasts generated by azacytidine treatment of 10T1/2 cells. They found that GH treatment induces the cells to form myotubes, whereas IGF-I does not exert any effect. Moreover, Adamafio et al (9) showed the presence of specific and functionally active GH receptors on mouse BC3H1 myoblasts. However, several experimental lines of evidence strongly suggest that IGFs (IGF-I, IGF-II) are the main contributors for myogenic differentation. Growth and development of muscle require the proliferation, differentation and fusion of myoblasts, and it is well documented that IGFs play a major role in controlling these processes (10). Moreover, IGFs are also potential candidates for regulation of satellite cell function during adult muscle regeneration (11).

METABOLIC ACTION OF GH ON SKELETAL MUSCLE

It is well known that GH exerts an important role in lipid, carbohydrate and protein muscle metabolism. Evidence has been provided that many of the metabolic actions of GH are mediated both directly through specific receptors for GH (12) and, indirectly, through the action of other hormonal factors, such as insulin and IGF-I which may act in either a hormonal or an autocrine paracrine fashion (13). There are substantial differences between the metabolic effects of GH and IGF-I. Indeed, IGF-I mediated effects may enhance or counteract the direct metabolic action of GH. Moreover, the metabolic effects of IGF-I are also regulated by related binding proteins, which may inhibit or enhance IGF-I effects, as well as muscle production of IGF-I and nutritional status. Tissue concentrations of IGF-I may be more important than plasma levels in exerting metabolic effects. ATP is the immediate chemical source of energy required by skeletal muscle. ATP synthesis is related to the oxidation of metabolic fuels, principally glucose, fatty acids, aminoacids and ketone bodies (acetoacetate and β-hydroxybutyrate). The relative use of these fuels depends on the individual's nutritional and hormonal status as well as exercise (14,15). In healthy subjects GH exerts profound effects on the utilizazion of fuels by muscle tissue.

PROTEIN METABOLISM

Role of GH

Muscle tissue accounts for 45-50% of the total body mass and 80-85% of fat-free mass, and for <30% of whole body protein synthesis (16). Protein metabolism largely contributes to the function and composition of skeletal musculature. GH can augment aminoacid uptake and increase protein synthesis in muscle, causing higher nitrogen retention and the preservation or increase in lean body mass. Several observations from animal investigations indicate that GH promotes protein anabolism. Experiments carried out on the rat have shown that GH stimulates the uptake of aminoacids into the diaphragm muscle (17) and increases the number and activity of ribosomes in muscle. Moreover, it enhances RNA polymerase (5). A 5 day treatment with GH in hypophysectomized rats induces an increase both in protein synthesis in the perfused hind-limb, and in muscle RNA (18). It is well known that GH is necessary for normal growth and development in childhood (19). However, GH deficency due to pituitary disease in adults results in reduced lean mass (20,21) and exercise capacity (22) which can be reversed by rhGH treatment. This demonstrates that an optimal level of GH is required for normal muscle mass and function and maintainance of protein balance. It has been shown that daily administration of rhGH in normal volunteers on caloric restriction (23) and in patients receiving parenteral nutrition (24) results in positive nitrogen balance. Protein synthesis rate, assessed using isotope-labeled aminoacids in post-operative patients treated with rhGH, is higher than that observed in placebo-controlled group (25). This protein anabolism results from a relative increase in protein synthesis compared to breakdown. Indeed, the direct action of GH on protein synthesis was deduced from observations that GH administration in humans during glucocorticoid treatment reduces the rate of aminoacid oxidation and increases that of aminoacid incorporation into proteins, resulting in increased whole-body protein synthesis (26,27). Several studies have proposed a stimulatory effect of GH on muscle protein synthesis (28-31). Adjuvant therapy with pharmacological doses of rhGH in post-operative patients has been found not only to increase whole-body nitrogen balance, but also to improve retention of glutamine as well to preserve total ribosome and polyribosome concentration of skeletal muscle, suggesting that

GH affects muscle composition (29). These findings are in agreement with the data of Lundeberg et al (32) who demonstrated that administration of rhGH reduces muscle protein catabolism and improves whole-body nitrogen economy in healthy male volunteers on hyponitrogenous oral diet. Clear evidence that GH is a potent anabolic agent promoting muscle protein synthesis specifically at muscle level and independently of its ability to increase circulating IGF1 levels, can be drawn from the studies of Fryburg et al, who demonstrated that local (28) or systemic (33) infusion of GH in healthy subjects acutely stimulates muscle, but not whole body protein synthesis, with negligible changes in systemic IGF-I concentrations. In addition, both limb aminoacid uptake and myosin heavy-chain mRNA are also increased by GH administration in nutritionally depleted humans (34), suggesting that GH can stimulate transcription of mRNA for myofibrillar proteins. The stimulative effect of GH on muscle protein metabolism has not, however, been confirmed by the findings of other investigators who failed to demonstrate any stimulatory effect of GH on muscle protein synthesis in spite of enhanced protein synthesis in the whole body (31,35). The GH-mediated increase in protein synthesis described in normal subjects has been confirmed in adult-GH deficient patients. When using isotopes in patients treated with rhGH for 2 months, it has been found, that GH does not affect proteolisis rates but directly stimulates whole-body protein synthesis (36). The mechanism of the anabolic action of GH has been reviewed by Russel-Jones and Umplety (37). GH shifts oxidation from proteins to fat utilization by promoting lipolysis, and stimulating aminoacid uptake and their incorporations into proteins (26,31,35,38). Therefore, in contrast to the decrease in protein breakdown, which is the mechanism controlled by insulin, the mechanism underlying GH-mediated protein anabolism is stimulation of protein synthesis. The anabolic action of GH administration for 3 days in normal subjects on whole body and muscular protein balance has been shown to be enhanced by co-administration of exogenous insulin (39).

Role of IGF-I

GH induces liver production of IGF-I (40) that is released into circulation until it arrives at target tissue, notably skeletal muscle (13). However, IGF-I is also produced in a wide variety of other tissues including muscle (41). Several studies have investigated the role of IGF-I in the protein metabolism of skeletal muscle. IGF-I increases muscle protein synthesis in "in vivo" animal models (42) and "in vitro" muscle cell cultures (43). By contrast, in hypophysectomized rats the increase in the weight of gastrocnemius and soleus muscles brought about by continuous subcutaneus infusion of IGF-I does not reach significance (44). Administration of rhIGF-I has been found to improve starvation-induced weight loss, enhance weight gain after burn injury and decrease protein degradation in fasting state (45,46). Studies in animals using isotopic tracers have shown that IGF-I reduces proteolisis but has no effect on protein synthesis (45,47). Jacob et al (45) demonstrated that infusion of IGF-I in fasted rats induces decreases in leucine turnover rate, incorporations of radiolabeled leucine into tissue protein and plasma aminoacid concentrations, indicating a reduction in protein breakdown. They concluded that the effect of IGF-I on protein metabolism is mediated primarily by inhibition of proteolysis. Even in septic rats IGF-I can reverse sepsis-induced inhibition of protein synthesis (48). In human studies it has been shown that infusion of IGF-I results in reduced proteolysis with no effect on protein synthesis, as demonstrated by the decrease in whole-body leucine flux and oxidation (49). It would seem, therefore, that the action mechanism of IGF-I is similar to that of insulin on protein metabolism. Similarly, subcutaneous infusion of IGF-I at a relatively high dose during a 5-day period, reduces protein oxidation in normal (50) and deficient subjects (51). However, when an adequate supply of substrate is given with an aminoacid clamp simultaneous infusion of IGF-I and aminoacids results in stimulation of protein synthesis in normal subjects, (52), which conflicts with observations from previous studies conducted in fasted subjects or animals. Accordingly, high doses of IGF-I increase lean body mass in elderly women to a similar degree as during GH administration (53). Data available in humans indicate that IGF-I may also exert an anabolic action on skeletal muscle. Fryburg et al (30) demonstrated that intra-arterial infusion of IGF-I promotes forearm tissue net anabolism by a 70% increase in protein synthesis. Interestingly, a recent study (54) indicated that IGF-I increases forearm muscle blood flow through a nitric oxide dependent mechanism, and that nitric oxide may be required for the protein synthetic response of muscle to IGF-I. To date it still remains to be elucidated whether GH-stimulated production of IGF-I is responsible for the muscle protein anabolic effect observed during GH administration. There are, in fact, some differences between GH and IGF-I in promoting muscle protein anabolism. It has been proposed that GH may exert direct effects on muscle protein balance regardless of an increase in circulating levels of IGF-I (26). Furthermore, the finding that infusion of GH in healthy humans increases protein synthesis in forearm muscle, while circulating IGF-I levels remain unchanged, indicates that it may exert direct action, (28) although it cannot be excluded that locally produced IGF-I can enhance the anabolic effects of GH. In addition, GH may contribute to improve whole body protein synthesis and preserve hand grip strength in post-operative patients, regardless of its ability to increase IGF-1 levels (55). When compared with GH's ability to stimulate growth of different organs in hypophysectomized rats, rhIGF-I has been found less effective than rhGH in increasing the weight of the gastrocnemius and soleus muscles (44). Studies in hypophysectomized rats have shown, however, that equipotent trophic effects cannot be attributed to IGF-I and GH since higher doses of IGF-I are necessary to obtain responses which are equivalent to those determined by rhGH (56). On the other hand, these data should also be explained in the light of recent observations on local production of IGF-I in the muscle (57). In fact, an activation of IGF expression has recently been shown during exercise-induced skeletal muscle hypertrophy in hypophysectomized rats (58). This, therefore, suggests that exercise may induce an increase in muscle IGFmRNA independently of GH. In agreement with these finding, muscle protein synthesis in rats after prolonged exercise has been found to be associated with an increase in muscle IGF-1 protein, whereas circulating IGF-I levels remain unchanged (59). Nevetheless, it has recently been shown that in GH-deficient young adults exercise alone, without exogenous GH supplementation, is not sufficient to maintain muscle mass and strength (60). Since GH exerts a potent anabolic action on protein metabolism, considerable time is being spent on investigating its use in the treatment of pathophysiological processes associated with protein hypercatabolism and loss of lean body mass, as in the case with burns, fractures, chronic lung, kidney, and liver disease, organic GH deficency, sepsis, AIDS, surgical stress, malnutrition, cachexia and normal aging.

GLUCOSE AND LIPID METABOLISM

Role of GH

Human muscles in adults markedly influence the rate of glucose utilization (61). Glucose taken up by muscle cells is either metabolized or stored as glycogen (very little indeed, because muscle glycogen accounts for only 1.5 % of the wet weight (62). GH has been shown to alter carbohydrate metabolism by producing hyperglycaemia and insulin resistance. In rat skeletal muscle (hemidiaphragm) glucose uptake (63) and its conversion to glycogen (64) are increased after exposure to GH. These results in vitro, however, are at odds with studies in humans which demonstrate that forearm infusion of GH acutely inhibits muscle glucose uptake and enhances the release of free fat acids (FFA) from adipose tissue (65). An "early" insulin-like effect of GH has been inferred to occur in vivo, since administration of GH has been found to decrease post-absorptive splanchnic glucose production and enhance glucose uptake and utilization, with concomitant decrease in plasma glucose concentrations (66). However, several studies in humans have shown an inhibitory action of elevated plasma GH levels on the insulin-mediated peripheral glucose uptake (67,68) which worsens glucose tolerance. There is additional evidence that even moderate increases in GH levels reduce hepatic and peripheral insulin sensitivity and that the impairment of peripheral insulin sensitivity mainly resides in muscle (69). Gore et al (70) found that burned patients given rhGH during hyperinsulinaemic euglycaemic clamp have a decrease in glucose uptake and an inhibition of glucose oxidation. These findings indicate that GH may promote insulin resistance also in hypercatabolic patients, and this effect is caused by an impairment in glucose transport. The relative insulin resistance is, therefore, due to a direct effect of GH, independent of IGF-I, which has, instead, been shown to increase intracellular glucose transport, as well as the glucose-induced glucose uptake (71) on the liver and extrahepatic tissues. GH has been shown to increase post-prandial hepatic glucose release both by stimulating gluconeogenesis and glycogenolysis (72) and by reducing glucose utilization in muscle tissue. There is evidence that the impairment of muscle glucose utilization could be attributable to the inhibitory action of GH on glycogen synthesis and glucose oxidation (70,73). It is well documented that GH inhibits insulin-mediated activation of the glycogen synthase in skeletal muscle by a mechanism which is distal to insulin receptor binding and kinase activity (73). The GH-promoted decrease in oxidative and non-oxidative glucose metabolism is offset by increased lipid oxidation rate, which contributes to a reduction in protein breakdown and gluconeogenesis (73). GH is an important regulator of the release and oxidation of lipids (74). There is a solid body of evidence that GH acts directly to cause lipolysis, by stimulating the intracellular hormone-sensitive lipase in the adipocyte to release free fatty acids (FFA) and glycerol, (75) and increases lipid oxidation rates (76). It is well known that FFA are a major substrate for oxidation and energy production in skeletal muscle, and an overflow of acetyl coenzime A derived from FFA mobilized by GH leads to the formation of ketone bodies. Glycerol derived from fat breakdown enters the gluconeogenic pathway. Current evidence suggests that the GH-promoted lipolysis results in a decrease in glucose oxidation and protein breakdown and an increase in gluconeogenesis. Several human studies have shown that GH administration increases circulating levels of FFA and glycerol, revealing a clear stimulation of lipolysis (31, 76). Mjaaland et al (77) have found that GH decreases carbohydrate oxidation and enhances fat oxidation, FFA, glycerol and β-hydroxybutyrate concentrations in patients who had undergone abdominal surgery. This suggests that GH therapy in these patients oxidizes fat in preference to glucose for energy production. In skeletal muscle the presence of lipoprotein lipase (LPL) has been documented. LPL is an enzyme localized in the endothelium of the vessel wall where it hydrolizes triglycerides into chylomicrons and very-low-density lipoprotein (VLDL) supplying fatty acids to muscle tissue (78). In particular, it has been shown in mice that uptake of fatty acids to muscle cells is directly related to tissue LPL activity (79). Moreover, much evidence has indicated that in humans an increase in skeletal muscle LPL activity improves lipid metabolism by inducing a decrease in triglyceride levels and an enhancement of HDL cholesterol and insulin sensitivity (80). The regulation of LPL activity is complex, and several factors, such as nutritional and endocrine factors, are involved (78,80). The effect of GH on skeletal muscle

LPL activity in humans is still unclear. Animal studies have demonstrated that GH administration in hypophysectomized rats increases skeletal muscle LPL activity, which suggests that GH might reduce glucose utilization in muscle tissue in favour of increased lipid oxidation by activating LPL activity (81). The stimulatory effect of administered GH on muscle LPL activity does not seem to be mediated by IGF-I since it has been shown in hypophysectomized rats that IGF-I administration results in a decrease of adipose tissue LPL activity, whereas there is no effect on muscle tissue LPL activity (81). On the contrary, Richelsen et al (82) have found that 4 month treatment with GH for GH-deficiency results in a significant reduction in adipose tissue LPL activity and no appreciable effect on skeletal muscle LPL activity. This proves that GH, at least with an antagonistic action, is not involved in the regulation of muscle LPL activity in humans.

Role of IGF-I

IGF-I is structurally and biologically related to insulin (83). The cellular action of IGF-I requires its interaction with a specific cell-surface receptor (type I IGF receptor) which is structurally similar to the insulin receptor (84). The insulin and IGF-I receptor can bind both ligands, however, with approximately 10-fold more affinity than it does the related ligand (85). Hepatocytes (86) and adipocytes (87), classified as target tissues for insulin, lack functional receptors for IGF-I. Skeletal muscle cells have receptors for both IGF-I and insulin (88). The differences in ligand affinities and tissue expression of the receptors explain the different metabolic action between IGF-I and insulin. Indeed, the two hormones exert different effects on glucose metabolism. IGF-I, acting through its own receptor (88), directly stimulates glucose uptake by muscle by promoting GLUT4 translocation (89), whereas it is less effective than insulin in inhibiting hepatic glucose production (49). Several animals studies (45,46) have shown that IGF-I administration increases glucose transport, largely in skeletal muscle (90), promotes glycolysis and glycogenolysis, and reduces gluconeogenesis. There is also evidence that IGF-I acts to promote muscle glucose uptake even in insulin resistent septic rats (91). Administration of IGF-I in normal (50) and GH deficient (51) adults has been reported to induce a 10-20% increase in basal metabolic rate associated with increased lipolysis and reduced proteolysis in the presence of unchanged basal glucose metabolism. It has been proposed, therefore, that IGF-I directly decreases muscle protein breakdown and promotes lipid oxidation by reducing the anti-lipolitic effect of insulin. Accordingly, intravenous IGF-I infusion, at a dose which clearly increases glucose uptake, has been found not to suppress lipolysis (92). There is evidence that IGF-I has no direct effect on adipose tissue in vivo since adipocytes lack functional receptors for IGF-I (87) and may exert a direct inhibitory action on pancreatic secretion of insulin (93). A study by Fryburg et al (30) in humans indicated that intra-arterial infusion of IGF-I increases forearm glucose uptake, confirming the substantial stimulatory effect of IGF-I on glucose uptake in peripheral tissue. It has been proposed that the insulin-like effects of IGF-I are due only to the free peptide and not to protein bound IGF-I (94). In healthy subjects less than 5% of circulating IGF-I is free, the remainder being reversibly bound to a family of six high affinity binding proteins (IGFBP-1 to-6), which influences its half-life and receptor availability, by variously potentiating or inhibiting IGF-I effects (95). However, under normal conditions, it appears that IGF-I does not have an important role in the regulation of glucose metabolism since the level of free- IGF-I is too low to elicit definite insulin-like effects. During states of enhanced catabolism GH basal levels are generally increased, whilst circulating IGF-I levels fall (96). This has been documented in critical illness (97), trauma (98), burns (99) and starvation (100), and has been attributed to a resistance of the IGF-I axis to GH stimulation. The increase in GH induces insulin resistance and lipolysis raising blood glucose concentrations and liberating free fatty acids, whereas the fall in circulating IGF-I promotes muscle protein breakdown. Such changes may be of great benefit for patients by providing them with metabolic fuels, thus avoiding the hazards of hypoglycaemia. However, it has been found that muscle IGF-I mRNA from critically ill patients is preserved despite the low levels in circulating IGF-I (101).

GH DEFICIENCY AND SKELETAL MUSCLE

Most hypopituitary patients complain of general fatigue, low physical capacity and muscle weakness, although they receive adequate and stable replacement with cortisone, thyroxine and gonadal steroids, and are capable of carrying out normal daily activities (102). It is well documented that GH deficiency (GHD) is frequently associated with pituitary disease (103,104) and a large body of evidence has accumulated in the past decade to show that long-standing GH deficiency is associated with alterations such as impaired quality of life, reduced physical strength and work capacity, and decreased lean body mass, all of which appear to be reversible with GH replacement therapy in most patients. Several studies have been performed on changes in muscle structure and function in subjects with GHD, and on the effect of GH replacement therapy on the muscle system. However, still unclear is the extent to which both childhood- and adult-onset GHD results in muscle structural and funcional changes, and whether GH standard replacement therapy is associated with substantial beneficial effects on muscle mass and strength.

GHD and skeletal muscle in children

In children, GHD is frequently associated with a slow motor development and a reduction in muscle mass and strength (105-109), although shortness of stature is the most apparent abnormality in these patients. Regular conventional replacement therapy in GH deficient children, besides promoting longitudinal bone growth, has important beneficial effects on body composition as it increases lean body mass, body cell mass and limb mass cross-sectional area (106,107,110-113). It is known that discontinuation of hGH replacement therapy in young GHD patients who are reaching adult-hood results in a detectable decrease in quadriceps muscle size and muscle strength within a period of 12 months, suggesting an important role for GH in the maintenance of muscle integrity in young individuals (113). Recently, Brat et al (114) have shown, in untreated children with isolated GHD, that muscle performance of the upper and lower limbs, assessed by measurement of muscle isometric force and endurance, is significantly reduced compared to normal matched controls, and can be progressively restored by GH treatment, mainly in boys during puberty. Similarly, in children with Laron syndrome IGF-I deficiency is associated with a lowering in muscle performance. However, patients treated with IGF-I for 3-4 years show a less striking improvement in muscle function perhaps because of the more marked muscle underdevelopment in this disease (115). These observations confirm that primary or secondary IGF-I deficiency in children leads to an impairment in skeletal muscle performance that may be corrected or improved with replacement therapy. Leger et al (116) reported in a group of GHD patients ,Turner's syndrome, and intrauterine growth retardation prepubertal patients that 1 year hGH treatment induces a marked increase in muscle tissue mass of the thighs, as measured by magnetic resonance imaging (MRI), with significant difference from the small variations observed in normal children followed longitudinally as controls. Subsequently, (117) these results were furtherly confirmed by the same authors who demonstrated that a 3-year hGH treatment significantly increases muscle tissue mass, as evaluated by MRI in the thigh, in children born small for gestational age, compared to normal children followed longitudinally. This suggests a beneficial long-term effect of rhGH administration on muscle mass also in non GHD short children.

GHD and skeletal muscle in adults

Several studies have shown that in adults, both child-hood and adult-onset GHD is associated with changes in skeletal musculature including decreased muscle mass and decreased muscle strength. (20,113,118,119,120-127). Various methods were used in the evaluation of muscle mass and muscle strength in untreated GH deficient adults. However, despite methodological differences, the results of these studies are quite similar. Assessement of total body protein potassium by 40K measurement in the adults with GHD showed that lean body mass, of which muscle is the major component, was 8% less than predicted for age, sex, height and weight (20). Mean muscle to fat area, estimated from the mid-thigh by computerized tomography (CT), as well as maximal voluntary isometric strength of quadriceps were found to be significantly reduced in a group of subjects with chilhood-onset GHD (118, 120). Accordingly, Cuneo et al (119, 121) reported that mid-thigh muscle cross-sectional area (CSA) (assessed by CT and expressed per kilogram body weight) as well as isometric quadriceps muscle strength (expressed per muscle unit) were significantly lower in the adult with long-standing severe GHD. However, isometric strength in muscle groups other than the quadriceps has shown practically no difference from a normal population. Significant decrease in quadriceps muscle CSA and strength were observed in an uncontrolled study of GH deficient young patients after the withdrawal of GH treatment for 12 months (113). Taken together these data indicate that adults with GHD have decreased muscle mass in comparison to normal individuals. On the contrary, uncertainty still remains regarding the reduction in muscle strength, which appears to be more related to the reduction in muscle mass than to intrinsic impairment of muscle contractile properties. Accordingly, Sartorio et al (124) have reported that patients with childhood-onset GHD, despite an absolute reduction in skeletal muscle mass and strength, have normal quadriceps strength per thigh muscle mass, as evaluated by anthropometry. Bottinelli et al (127) investigated contractile properties and fiber type distribution of quadriceps muscles in patients with childhood-onset GHD. They reported no significant difference between GHD patients and healthy controls in quadriceps muscle twitch characteristics, fatigue index and fiber type distribution. Similarly, no significant difference was found between patients and controls in both muscle strength and fiber CSA normalized over quadriceps CSA and body height, respectively, although the absolute values of both strength and fiber CSA were significantly lower in GHD subjects. These findings indicate that the reduction in muscle CSA and body size may account for the decline in skeletal muscle function and fiber size in GHD patients, and suggest that intrinsic muscle strength in these patients is not significantly changed. Therefore, muscular complaints in patients with childhood-onset GHD should not simply be attributed to abnormalities related to the structure or function of skeletal muscles. Janssen et al (126) recently confirmed these observations in adults with GHD as they found a significant decrease in quadriceps muscle mass of adults with GHD, as measured by MRI, and in isometric muscle strength as well as a small reduction in isokinetic muscle strength compared to age- and height-matched controls. However, when strength was expressed per muscle mass unit, no significant differences were found between GHD patients and healthy controls, confirming the concept that reduction in muscle strength is predominantly related to reduction in muscle mass. Similarly, muscle bioenergetic state, assessed by phosphorus nuclear magnetic resonance spectroscopy was not found to be significantly different in adults with GHD as compared to that in healthy subjects. In conclusion, GH deficient adults have moderately decreased muscle strength which can mainly be attributed to the reduction in muscle mass which is known to occur frequently in GHD.

GHD AND HISTOLOGY OF SKELETAL MUSCLE

Histological examination of muscle biopsies taken from vastus lateralis in adults with GHD have shown no changes in fiber size and distribution in comparison with healthy subjects (128,129). This indicates that abnormal muscle fiber density or proportion may not account for the fatigability demonstrated in adults with GHD, although more studies are necessary before definitive conclusions can be drawn regarding fiber type proportions and areas. Skeletal muscle fibers have been classified into slow twitch red (type 1) and fast twitch white (type 2) based upon their myoglobin content and time-to-peak tension (130). Type 1 fibers have a rich capillary blood supply, high mitochondrial and low fibrillar ATPase content, and low glycolitic and high oxidative capacity, are fatigue resistent and recruited for low-level contractions via aerobic metabolism. Type 2 fibers have a poor capillary blood supply and a high glycolytic but low oxidative capacity, and are fatigue susceptible and recruited for short bursts of rapid, heavy work with energy supplied by anaerobic glycolysis. Muscle type 2 fibers have been further divided into the subtypes of 2A and 2B with type 2A fibers displaying much higher oxidative capacity and lower fatigability than type 2B fibers. In four of five young GHD patients, muscle fiber atrophy associated with a decrease in fiber size ranging from 10-40% has been documented 1 year after withdrawal of GH treatment, even if no changes in fiber type composition have been shown (113). Recently patients with GHD are reported as having short half-relaxation time and rightward shift of force-frequency relation of quadriceps muscle compared to controls which provides indirect evidence for an increased proportion of more easily fatigable type 2 fibers in hypopituitary patients (125). These observations, thus, might suggest that GH can influence fiber type distribution in adults and do not rule out the possibility that alterations in strength generation could be due to type 2 fiber predominance. However, previous studies showing normal type 1 and type 2 fiber areas and proportions (128,129) have been confirmed by recent data of Bottinelli et al, (127) who did not find any difference between GHD patients and controls as far as fiber type distribution of vastus lateralis muscle was concerned as estimated by determination of myosin heavy chain (MHC) isoform distribution.

SKELETAL MUSCLE IN ACROMEGALY

In acromegaly, the muscles often appear hypertrophied but are functionally impaired and weaker. The presence of enlarged muscles associated with weakness and hypotonia in patients with acromegaly was first reported by Pierre Marie (131). These patients showed increased muscle bulk and strength early in the disease, and then became increasingly weak and fatigable as the disease progressed. Subsequently, Souza-Leite (132,133) reported decreased exercise tolerance in acromegaly in 1891. Despite these observations, there have been few clinical reports on skeletal muscle alterations in patients suffering from chronic GH hypersecretion. More recent studies (134) and Lundberg (135) have revealed that proximal miopathy occurs in 50% of acromegalic patients. Pickett et al (136) have confirmed that in acromegalic patients chronic GH excess is associated with mild proximal weakness and flabbiness of muscles which are insidious at the beginning but gradually progress. Myopathic changes in these patients correlated with the duration of their underlying disease: the longer a patient was exposed to high concentrations of GH, the more he was likely to develop myopathy. Moreover, the electromyography in these patients usually showed a myopathic pattern with short-duration, low-amplitude motor unit potentials and increased frequency of polyphasic potentials with no other distinctive features (136). The observation that patients with acromegaly have larger but not stronger muscles is in line with results of animal studies demonstrating that the increase in muscle size and weight occurring during GH excess is not associated with a proportionate increase in performance. Greenbound and Young (137) reported that GH administration in rats leads to an increase in muscle bulk which in proximal muscles, such as pectoralis major, vastus lateralis, supraspinatus, and deep muscles of the spine, is found to be higher than that in distal muscles such as the gastocnemius, soleus, and tibialis anterior. Accordingly, Bigland and Jehring (4) later showed that a 3 week GH treatment in rats results in an increase of both quadriceps muscles by 15-40%, and cross-sectional area of the myofibrils by 6-20%; however, the performance of the hypertrophied muscles was not affected and their contractile strength on the basis of weight was lower than that of normal muscles. Recently, Wolf et al (138) have reported that transgenic mice expressing bovine GH and exposed to permanently elevated GH levels exibit an increase in muscle weight which is not associated with a proportionate increase in muscle strength. These findings are in agreement with recent data of Costa et al (139) indicating that transgenic rabbits expressing bovine GH have an enlargment of the limbs with histological alterations which closely resemble those found in acromegalic patients. In fact, it cannnot be excluded that the muscular alterations seen during GH hypersecretion might depend either on muscle oedema associated with GH excess or on GH-induced changes in muscle fiber type distribution with a relative enhancement of slow type 1 fibers. Histological studies in rats bearing GH3 cell tumors have shown that GH excess is associated with a significant increase in type 1 and 2B muscle fibers with no change in type 2A fibers (140). Moreover, GH excess in hypophysectomized rats has been found to increase the proportion of type 1 fibers in skeletal muscles by conversion of transitional type 2/1 fibers into type 1 fibers (141), whereas an increase in type 2 muscle fiber relative content has been shown in GH-deficient rats (142). The available data in humans have shown that histological abnormalities associated with GH excess are not severe and are not sufficient to account for the clinical weakness (134,136,143,144). Muscle biopsies from acromegalic patients have shown selective hypertrophy of the type 1 fibers, whereas type 2 fibers are atrophic (145). However, Mastaglia (144) reported no consistent pattern of muscle fiber type hypertrophy or atrophy in acromegalic patients. Electron microscopy has revealed accumulation of glycogen, lipofuscin pigment and phospholipids as well as mitocondrial changes, indicative of a non-specific myopathy (136). A slow improvement in muscle strength has been shown during the first year after hypophysectomy in some patients, with no relationship however to post-operative decline in GH levels (136).

EFFECTS OF GH TREATMENT ON SKELETAL MUSCLE IN GHD

The advent of recombinant technology has greatly increased the availability of recombinant human growth hormone (rhGH), and has made the utilization of this hormone possible in terms of physiological replacement in GHD state. As a result greater insight into the physiological action of GH in adult life has been gained. There is evidence indicating that rhGH replacement therapy can reverse many of the abnormalities in the body composition, cardiovascular function, lipid profile and psychological well-being which are frequently reported in GHD patients (20,119,121,146-152). However, knowledge of the long-term effects of rhGH replacement therapy on skeletal muscle in GHD patients is limited, and whether treatment of GHD results in a sustained improvement also with regard to muscle mass and muscle strength is still a controversial issue. As earlier reported (vide supra) the withdrawal of GH administration in young GHD adults resulted in a reduction in maximal voluntary isometric muscle strength, muscle size and muscle fiber area (113), confirming the important role of GH in maintaining muscle morphology and function in adults.

EFFECT OF GH ON MUSCLE MASS

Placebo-controlled studies have reported that administration of rhGH to GHD adults for a period of 4-12 months leads to an increase in thigh muscle mass of approximately 5-11% (118,121,149,150,153). Moreover, muscle mass continues to recover for as long as GH replacement therapy continues. In fact, prolonged rhGH replacement in GHD adults has been shown to induce a progressive increase in leg muscle mass which exceeds baseline value by 28.7% after 5 years of treatment (154). Additionally, it has recently been reported that adult GHD patients receiving constant rhGH replacement therapy for 10 years show a significant increase in the total cross-sectional muscle area of the dominant thigh as measured by CT, whereas there is no change in untreated GHD patients (155). Overall, these studies confirm that long-term rhGH replacement therapy in GHD adults results in an apparent gain in muscle mass. It cannot be excluded that the increase in muscle volume detected sharply after initiation of GH treatment may, in part, be due to muscular fluid accumulation (156). No effects on skeletal muscle fiber size and proportion of type 1 vs 2 fibers have been documented after 6 months of rhGH therapy (128,129).

EFFECT OF GH ON MUSCLE STRENGTH

Several studies have addressed the effects of GH replacement therapy on muscle strength in GHD patients by testing isometric and isokinetic muscle strength in various muscle groups. Short-term placebo controlled (118,121,150) and uncontrolled (102,157) studies have demonstrated that rhGH treatment for 4 to 12 months does not significantly improve muscle strength. One controlled study has shown that limb girdle strength increases after 6 months of GH treatmant (121). Nevertheless, longer-term (up to 4 years) uncontrolled studies have shown an improvement in muscle after prolonged GH treatment (123,158,159), suggesting that the beneficial effects of GH therapy on muscle function in GHD adults are time-dependent. Recently, Wallymahmed et al (160) reported in GHD adults that muscle strength, evaluated by measuring maximum voluntary contraction at quadriceps level, does not change during the first year of rhGH treatment, but improves significantly during the second year of therapy, confirming that longer-term therapy may be needed to increase skeletal muscle strength. It has been speculated that the beneficial effects of long-term rhGH administration on muscle might be related to an increased physical activity associated with the improvement in body composition, feeling of well-being and vitality which occurs after constant long-term GH replacement. In fact, rhGH replacement can overcome the low energy levels and mental fatigue experienced by GHD patients who generally adapt to a sedentary lifestyle, which may contribute to low muscle mass and muscle strength (149,151, 152). It has therefore, been proposed that in patients treated with rhGH for long periods the enhanced exercise capacity may lead to a concomitant improvement in muscle function. This assumption is in line with the work of Goldberg (161) who reported that work-induced muscle growth may not be pituitary GH dependent. Accordingly, recent studies in rats have demonstrated that skeletal muscle IGF-I mRNA is activated by synergistic overload (58), treadmill running (162), and stretch-induced hypertrophy (163), and that this response is independent of GH status. Exercise performance and capacity are compromised in GHD adults and have been shown to improve after GH treatment in some (118,121,150,153), but not all studies (102,152,164). The improvement in physical capacity associated with rhGH therapy seems to be due to an increase in lean body mass and muscle mass combined with the enhanced cardiac performance. Recently, Gibney et al (155), when comparing muscle strength between two matched groups of GHD adults, demonstrated that the sum of muscle strength as tested in three different muscle groups (elbow, shoulder and hip) significantly declines in the group of untreated GHD, but not in the group of GHD patients receiving uninterrupted rhGH replacement for 10 years. This confirms that restoration of normal levels of GH for a long period of time may have a beneficial effect on muscle function. The GH-mediated improvement in muscle strength can be assumed to be due to the increase in muscle cross-section area associated with GH administration (159). Accordingly, a recent study that has specifically addressed this question by investigating intrinsic muscle strength (strength expressed per muscle volume unit) in GHD adults, has reported a significant increase in quadriceps muscles volume and muscle strength with no change in intrinsic muscle strength, indicating that the increase in muscle strength can be attributed to the increase in muscle mass (126). Although general endurance as assessed on bicycle ergometers has been shown to increase in response to GH treatment (22,164) local fatigue index as tested at shoulder (121) and quadriceps muscles (159) has not been found to improve after 6 and 24 months of rhGH treatment, respectively. In conclusion, current findings may support the assumption that long-term GH replacement in GHD adults improves muscle strength, which has mainly to be attributed to the increase in muscle mass.

AGING AND SKELETAL MUSCLE

Normal aging is associated with marked changes in body composition, skeletal muscle size and function, and energy metabolism. These changes include an increase in total and visceral fat mass and a reduction in fat-free mass (primarily due to a loss of skeletal muscle mass), muscle strength and power, and resting metabolic rate (165-169). There is a large body of evidence to indicate that muscle strength declines with aging. Muscle strength reaches peak value between 25 and 35 yrs, approaches a plateau or is slightly lower between 40 and 49 yrs, and then decreases by approximately 12-14% per decade after 50 yrs of age (167,168). In addition, in 60-70 year-old individuals the relaxation after twitch contraction from the ankle dorsiflexor, plantar flexor, and thenar muscle groups have been reported to be 20-43% slower than that in young adults (170-172). The age associated differences in strength are related with age differences in muscle mass (173). A high correlation between muscle strength and muscle mass has been reported in healthy elderly individuals (174). Muscle mass has been shown to decrease by 35-40% from about 20 to 80 years of age (175,176) and these changes have been recognized to appear earlier in men (between 41 and 60 yrs) than in women (after 60 yrs) (177). The age-related decline in skeletal muscle has been attributed to a reduction in both muscle cell mass and muscle fiber number. In 70-73 year-old men, the number of muscle fibers in the vastus lateralis has been reported to be 23% less than that in young men aged 19-37 yrs (178). This reduction occurs largely in fast-twitch, type 2 fibers which decreases from 60 to 30% of muscle with increasing age (178,179). Moreover, evidence of reduced myofibrillar protein system rates in the elderly (180) has been reported; these changes may therefore contribute to further the alterations in muscle structure and function associated with aging in humans. The age-related loss of muscular mass and function has health and economic implications, because it is associated with a higher rate of functional and physical disabilities in the elderly (181). At the present time, the underlying causes responsible for the decline in skeletal muscle mass and strength observed during normal aging are not well understood. Nevertheless, one mechanism which may contribute to these changes is a declining activity of GH-IGF-I axis with increasing age. This decline in healthy humans has been well established (182), and there is evidence suggesting that it might contribute to the age-related changes in muscular function (183,184). To date, it is well known that organic GHD is associated with decrease in protein synthetic capacity and lean body mass and an expansion of fat mass (20,118,119,121,185). In spite of the lack of unequivocal evidence showing a casual relationship between age-related GH and IGF-I deficit and physical alterations, the findings that adverse changes in body composition of organic GH deficiency are strikingly similar to those of normal aging and can be reversed in GH-deficient patients by GH replacement (20,118,121,151) have led to the rationale for treating healthy elderly subjects with rhGH to improve skeletal muscle mass and strength. An increased interest has, therefore, developed regarding the potential clinical use of GH in the elderly in order to prevent, reverse or attenuate age-related alterations in muscle function and performance. The first study of GH replacement, performed in rodents, showed that GH increases tissue protein synthetic capacity in old animals to the levels normally found in young animals (186). In human studies, the administration of rhGH in elderly healthy subjects has generally been reported to promote an increase in IGF-I levels, total lean mass and muscle mass and strength (187-190). Administration of rhGH for 1 week in three different dosages induced a dose-dependent increase in IGF-I and a reduction in daily urinary nitrogen, phosphate and sodium excretion in twenty healthy men and women aged 60 yrs and over (187). Strong evidence that GH replacement therapy in elderly individuals can reverse some of the age-related alterations has been provided by Rudman et al (188) who found that 6 month rhGH treatment in healthy men, aged 61-81 yrs, selected to have lower (<0.35 U/ml) than normal age-adjusted IGF-I level, produced an 8.8% increase in lean body mass and a 14.4% loss in fat mass as compared with a group of 9 men receiving only placebo. Muscle strength, measured as maximum torque generated on an isokinetic dynamometer, improved by 38% for knee extention and 23% for knee flexion in a group of healthy elderly men treated with rhGH for 3 months (0.03 mg/kg, sc, 3x/week), but whole-body protein breakdown or synthesis and myofibrillar protein synthesis in the quadriceps were unaltered, indicating that GH supplementation cannot restore myofibrillar protein synthesis to such an extent as to be detected in young people (189). By contrast, Butterfield et al (190) have shown that administration of rhGH for 1 month results in a 50% increase in mixed muscle protein synthesis. However, other studies in humans have demonstrated that exercise can improve muscle function to the same extent as rhGH therapy and at far less cost (191). Taaffe et al (192,193) have shown that in healthy elderly men admistration of rhGH (0.02 mg/Kg per day for 10 weeks) failed to increase muscle size and strength over what could be obtained with 24 weeks of regular resistance exercise programmes alone. The results of the previous study are similar to the observations that in healthy elderly subjects with low levels of GH, muscle atrophy and weakness GH treatment results in an increase in fat free mass, muscle strength and whole body protein synthesis only if associated with a 12-to 16 week resistance exercise training (35,191). Recently rhGH administration in elderly men has been reported not to increase knee flexion or extention, hand grip strength or systemic endurance (194) This and preceeding studies suggest that replacement GH therapy in healthy elderly subjects does not reverse the age-related decline in muscle strength. Despite some promising observations, the long-term GH supplementation in healthy elderly subjects with low IGF-I levels can be associated with sodium and fluid retention, carpal tunnel syndrome, glucose intolerance, hyperinsulinemia and gynecomastia (20). The incidence of these long- term side effects necessitates careful examination and appears to be related to the dose of GH given, the level of IGF-I response and duration of treatment. These complications may be reduced by adjusting the GH dosage to induce optimum IGF-I levels. Moreover, although the side effects disappear when the GH dosage is reduced, the long-term safety of GH suplementations in elderly subjects without pituitary disease still remains to be established, particularly with respect to the risk of cardiovascular disease and tumor initiation and progression. In acromegalic patients an increased prevalence of hypertention and cardiomyopathy has been found as well as an increase in acute cardiovascular and cerebrovascular attacks (195,196). In addition, the mitogenic action of GH and IGF-I has been well documented. GH administration stimulates c-myc proto-oncogene expression in liver and kidney of hypophysectomized rats (197). Acromegalic patients exhibit an increased risk of malignant diseases, of which colorectal and breast cancer are the most frequent (for review, see 198). A recent report has indicated a link between IGF-I levels and the incidence of prostate cancer (199). Taken together, the data available indicate the potential advantages of GH administration on body composition and, to a lesser extent, muscle mass and strength in healthy elderly subjects with declining GH secretion. This could therefore provide a rationale for therapeutic use of rhGH replacement in the elderly to reverse , in part, the adverse age-related changes. However, since the clinical efficacy of such an approach is modest and the incidence of long-term side effects and financial costs associated with rhGH treatment are high, GH supplementation in healthy elderly individuals with declining GH secretion cannot, at present, be justified to attenuate age-related alterations in muscle function and performance. Until more studies are available to establish the safety and efficacy of GH administration in elderly subjects without pituitary disease, but with clinical features resembling those of GHD, it is best if rhGH replacement is restricted to patients with organic GHD syndrome. It may be more advisable to encourage healthy elderly subjects to preserve the skeletal muscle mass and strength by adequate nutritional and exercise programmes. In conclusion, the clinical utility of replacement dose of GH to improve muscle performance in elderly subjects still remains to be demonstrated. Further experimental data and experience are warranted to explore the anabolic application of GH in diseases associated with progressive decrease in skeletal muscle mass and strength.

REFERENCES

1. Florini JR, Ewton DZ, Coolican SA. Growth hormone and the insulin-like growth factor system in myogenesis. Endocr Rev 1996;17:481-517.
2. Lee MO, Schaffer NK. Anterior pituitary growth hormone and the composition of growth. J Clin Nature 1934;7:337-367.
3. Plattener EB, Reed CI. A study of muscular efficency in rats injected with anterior pituitary growth factor. Endocrinology 1939;25:401.
4. Bigland B, Jehring B. Muscle performance in rats, normal and treated with growth hormone. J Physiol 1952;116:129.
5. Florini JR, Breuer CB. Amino acid incorporation into protein by cell-free system from rat skeletal muscle. Effects of pituitary growth hormone on activity of ribosomes and ribonucleic acid and RNA polymerase in hypophysectomized rats. Biochem 1966;5:1870-1876.
6. Palmer RM, Loveridge N, Thomson BM, Mackie SC, Tonner E, Flint DJ. Effects of a polyclonal antiserum to rat growth hormone on circulating insulin-like growth factor

(IGF)-I and IGF-binding protein concentrations and the growth of muscle and bone. J Endocrinol 1994;142:85-91.

7. Loughna PT, Bates PC. Interactions between growth hormone and nutrition in hypophysectomised rats:skeletal muscle myosin heavy chain mRNA levels. Biochem Biophys Res Commun 1994;198:97-102.
8. Nixon BT, Green H. Growth hormone promotes the differentiation of myoblasts and preadipocytes generated by azacytidine treatment of 10T1/2 cells. Proc Natl Acad Sci 1984;81:3429-3432.
9. Adamafio NA, Towns RJ, Kostyo JL. Growth hormone receptors and action in BC3H-1 myocytes. Growth Regul 1991;1:17-22.
10. Florini JR, Ewton DZ, Magri KA. Hormones, growth factors, and myogenic differentiation. Ann Rev Physiol 1991;53:201-16.
11. Edwal D, Schalling M, Jennische E, Norstedt G. Induction of insulin-like growth factor I messenger ribonucleic acid during regeneration of skeletal muscle. Endocrinology 1984;124:820-825
12. Oakes SR, Haynes KM, Waters MJ, Herington AC, Werther GA. Demonstration and localization of growth hormone receptor in human skin and skin fibroblasts. J Clin Endocrinol Metab 1992;75:1368-1373.
13. Isgaard J, Nilsson A, Vikman K, Isaksson OG. Growth hormone regulates the level of insulin-like growth factor-I mRNA in rat skeletal muscle. J Endocrinol 1989;120:107-112.
14. Newsholme EA, Leech AR. Biochemistry for the medical sciences. New York; John Wiley& Sons 1983.
15. Ruderman NB, Goodman MN, Conover CA, Berger M. Substrate utilization in perfused skeletal muscle. Diabetes 1979;28:13-17.
16. Balagopal P, Rooyackers OE, Adey DB, Ades PA, Nair KS. Effects of aging on in vivo synthesis of skeletal muscle myosin heavy-chain and sarcoplasmic protein in humans. Am J Physiol 1997;273:E790-E800.
17. Kostyo JL. Separation of the effects of growth hormone on muscle amino acid transport and protein synthesis. Endocrinology 1968;75: 113-119.
18. Flaim KE, Li JB, Jefferson LS. Protein turnover in rat skeletal muscle: effects of hypophysectomy and growth hormone. Am J Physiol 1978;234:E38-E43.
19. Aschner B. Uber die funktion der hypophyse. Pflugers Archiv 1912;146:1.
20. Salomon F, Cuneo RC, Hesp R, Sonksen PH. The effects of treatment with recombinant human growth hormone on body composition and metabolism in adults with growth hormone deficiency. N Engl J Med 1989;321:1797-803.
21. De Boer H, Blok GJ, Voerman HJ, De Vries PM, Van Der Veen EA. Body composition in adult growth hormone-deficient men, assessed by anthropometry and bioimpedance analysis. J Clin Endocrinol Metab 1992;75:833-837.
22. Cuneo RC, Salomon F, Wiles CM, Hesp R, Sonksen PH. Growth hormone treatment in growth hormone-deficient adults. II. Effects on exercise performance. J Appl Physiol 1991;70:695-700.
23. Kupfer SR, Underwood LE, Baxter RC, Clemmons DR. Enhancement of the anabolic effects of growth hormone and insulin-like growth factor I by use of both agents simultaneously. J Clin Invest 1993;91:391-396.
24. Ziegler TR, Young LS, Manson JM, Wilmore DW. Metabolic effects of recombinant human growth hormone in patients receiving parenteral nutrition. Ann Surg 1988;208:6-16.

25. Ward HC, Halliday D, Sim AJ. Protein and energy metabolism with biosynthetic human growth hormone after gastrointestinal surgery. Ann Surg 1987;206:56-61.
26. Horber FF, Haymond MW. Human growth hormone prevents the protein catabolic side effects of prednisone in humans. J Clin Invest 1990;86:265-72.
27. Bennet WM, Haymond MW. Growth hormone and lean tissue catabolism during long-term glucocorticoid treatment. Clin Endocrinol (Oxf) 1992;36:161-164.
28. Fryburg DA, Gelfand RA, Barrett EJ. Growth hormone acutely stimulates forearm muscle protein synthesis in normal humans. Am J Physiol 1991;260:E499-E504.
29. Hammarqvist F, Stromberg C, Von Der Decken A, Vinnars E, Wernerman. Biosynthetic human growth hormone preserves both muscle protein synthesis and the decrease in muscle-free glutamine, and improves whole-body nitrogen economy after operation. Ann Surg 1992;216:184-191.
30. Fryburg DA. Insulin-like growth factor I exerts growth hormone- and insulin-like actions on human muscle protein metabolism. Am J Physiol 1994;267:E331-336.
31. Copeland KC, Nair KS. Acute growth hormone effects on amino acid and lipid metabolism. J Clin Endocrinol Metab 1994;78:1040-1047.
32. Lundeberg S, Belfrage M, Wernerman J, Von der Decken A, Thunell S, Vinnars E. Growth hormone improves muscle protein metabolism and whole body nitrogen economy in man during a hyponitrogenous diet. Metabolism 1991;40:315-322.
33. Fryburg DA, Barrett EJ. Growth hormone acutely stimulates skeletal muscle but not whole-body protein synthesis in humans. Metabolism 1993;42:1223-1227.
34. Fong Y, Rosenbaum M, Tracey KJ, et al. Recombinant growth hormone enhances muscle myosin heavy-chain mRNA accumulation and amino acid accrual in humans. Proc Natl Acad Sci 1989;86:3371-3374.
35. Yarasheski KE, Campbell JA, Smith K, Rennie MJ, Holloszy JO, Bier DM. Effect of growth hormone and resistance exercise on muscle growth in young men. Am J Physiol 1992;262:E261-E267.
36. Russell-Jones DL, Weissberger AJ, Bowes SB, et al. The effects of growth hormone on protein metabolism in adult growth hormone deficient patients. Clin Endocrinol (Oxf) 1993;38:427-431.
37. Russell-Jones DL, Umpleby M. Protein anabolic action of insulin, growth hormone and insulin-like growth factor I. Eur J Endocrinol 1996;135:631-642.
38. Douglas RG, Humberstone DA, Haystead A, Shaw JH. Metabolic effects of recombinant human growth hormone: isotopic studies in the postabsorptive state and during total parenteral nutrition. Br J Surg 1990;77:785-90.
39. Wolf RF, Heslin MJ, Newman E, Pearlstone DB, Gonenne A, Brennan MF. Growth hormone and insulin combine to improve whole-body and skeletal muscle protein kinetics. Surgery 1992;112:284-291.
40. Schwander JC, Hauri C, Zapf J, Froesch ER. Synthesis and secretion of insulin-like growth factor and its binding protein by the perfused rat liver: dependence on growth hormone status. Endocrinology 1983;113:297-305.
41. D'Ercole AJ, Stiles AD, Underwood LE. Tissue concentrations of somatomedin C: further evidence for multiple sites of synthesis and paracrine or autocrine mechanisms of action. Proc Natl Acad Sci 1984;81:935-939.
42. Svanberg E, Adams S, Rosen D, Lundholm K. The role of IGF/IGFBP3 complex in regulation of muscle protein synthesis. Clin Nutr (Edinb) 1996; (Suppl. 1/15):26.

43. Roeder RA, Hossner KL, Sasser RG, Gunn JM. Regulation of protein turnover by recombinant human insulin-like growth factor-I in L6 myotube cultures. Horm Metab Res 1988;20:698-700.
44. Guler HP, Zapf J, Scheiwiller E, Froesch ER. Recombinant human insulin-like growth factor I stimulates growth and has distinct effects on organ size in hypophysectomized rats. Proc Natl Acad Sci 1988;85:4889-4893.
45. Jacob R, Barrett E, Plewe G, Fagin KD, Sherwin RS. Acute effects of insulin-like growth factor I on glucose and amino acid metabolism in the awake fasted rat. Comparison with insulin. J Clin Invest 1989;83:1717-1723.
46. O'Sullivan U, Gluckman PD, Breier BH, Woodall S, Siddiqui RA, McCutcheon SN. Insulin-like growth factor-1 (IGF-1) in mice reduces weight loss during starvation. Endocrinology 1989;125:2793-2794.
47. Umpleby AM, Shojaee-Moradie F, Thomason MJ, et al. Effects of insulin-like growth factor-I (IGF-I), insulin and combined IGF-I-insulin infusions on protein metabolism in dogs. Eur J Clin Invest 1994;24:337-344.
48. Jurasinski CV, Vary TC. Insulin-like growth factor I accelerates protein synthesis in skeletal muscle during sepsis. Am J Physiol 1995;269:E977-E981.
49. Laager R, Ninnis R, Keller U. Comparison of the effects of recombinant human insulin-like growth factor-I and insulin on glucose and leucine kinetics in humans. J Clin Invest 1993;92:1903-1909.
50. Hussain MA, Schmitz O, Mengel A, et al. Insulin-like growth factor I stimulates lipid oxidation, reduces protein oxidation, and enhances insulin sensitivity in humans. J Clin Invest 1993;92:2249-56.
51. Hussain MA, Schmitz O, Mengel A, et al. Comparison of the effects of growth hormone and insulin-like growth factor I on substrate oxidation and on insulin sensitivity in growth hormone-deficient humans. J Clin Invest 1994;94:1126-1133.
52. Russell-Jones DL, Umpleby AM, Hennessy TR, et al. Use of a leucine clamp to demonstrate that IGF-I actively stimulates protein synthesis in normal humans. Am J Physiol 1994;267:E591-E598.
53. Thompson JL, Butterfield GE, Marcus R, et al. The effects of recombinant human insulin-like growth factor-I and growth hormone on body composition in elderly women. J Clin Endocrinol Metab 1995;80:1845-1852.
54. Fryburg DA. NG-monomethyl-L-arginine inhibits the blood flow but not the insulin-like response of forearm muscle to IGF- I: possible role of nitric oxide in muscle protein synthesis. J Clin Invest 1996;97:1319-1328.
55. Jiang ZM, He GZ, Zhang SY, et al. Low-dose growth hormone and hypocaloric nutrition attenuate the protein-catabolic response after major operation. Ann Surg 1989;210:513-524.
56. Schoenle E, Zapf J, Hauri C, Steiner T, Froesch ER. Comparison of in vivo effects of insulin-like growth factors I and II and of growth hormone in hypophysectomized rats. Acta Endocrinol (Copenh) 1985;108:167-174.
57. Loughna PT, Mason P, Bates PC. Regulation of insulin-like growth factor 1 gene expression in skeletal muscle. Symp Soc Exp Biol 1992;46:319-330.
58. DeVol DL, Rotwein P, Sadow JL, Novakofski J, Bechtel PJ. Activation of insulin-like growth factor gene expression during work-induced skeletal muscle growth. Am J Physiol 1990;259:E89-95.

59. Eliakim A, Moromisato M, Moromisato D, Brasel JA, Roberts C Jr, Cooper DM. Increase in muscle IGF-I protein but not IGF-I mRNA after 5 days of endurance training in young rats. Am J Physiol 1997;273:R1557-R1561.
60. Sonksen PH. Growth hormone replacement therapy. In Savill J (ed) Horizons in Medicine Vol 9, 1997. London: Royal College of Physicians.
61. Wolfe RR, Allsop JR, Burke JF. Glucose metabolism in man: responses to intravenous glucose infusion. Metabolism 1979;28:210-220.
62. Saltin B, Henriksson J, Nygaard E, Andersen P, Jansson E. Fiber types and metabolic potentials of skeletal muscles in sedentary man and endurance runners. Ann N Y Acad Sci 1977;301:3-29.
63. Albertsson-Wikland K, Isaksson O. Time course of the effect of growth hormone in vitro on amino acid and monosaccharide transport and on protein synthesis in diaphragm of young normal rats. Endocrinology 1978;102:1445-1451.
64. Davidson MB. Effect of growth hormone on carbohydrate and lipid metabolism. Endocr Rev 1987;8:115-131.
65. Fineberg SE, Merimee TJ Acute metabolic effects of human growth hormone. Diabetes 1974;23:499-504.
66. MacGorman LR, Rizza RA, Gerich JE Physiological concentrations of growth hormone exert insulin-like and insulin antagonistic effects on both hepatic and extrahepatic tissues in man. J Clin Endocrinol Metab 1981;53:556-559.
67. Rizza RA, Mandarino LJ, Gerich JE. Effects of growth hormone on insulin action in man. Mechanisms of insulin resistance, impaired suppression of glucose production, and impaired stimulation of glucose utilization. Diabetes 1982;31:663-669.
68. Bratusch-Marrain PR, Smith D, DeFronzo RA. The effect of growth hormone on glucose metabolism and insulin secretion in man. J Clin Endocrinol Metab 1982;55:973-982.
69. Moller N, Butler PC, Antsiferov MA, Alberti KG. Effects of growth hormone on insulin sensitivity and forearm metabolism in normal man. Diabetologia 1989;32:105-110.
70. Gore DC, Honeycutt D, Jahoor F, Rutan T, Wolfe RR, Herndon DN. Effect of exogenous growth hormone on glucose utilization in burn patients. J Surg Res 1991;51:518-523.
71. Orskov L, Schmitz O, Jorgensen JO, et al. Influence of growth hormone on glucose-induced glucose uptake in normal men as assessed by the hyperglycemic clamp technique. J Clin Endocrinol Metab 1989;68:276-282.
72. Butler P, Kryshak E, Rizza R. Mechanism of growth hormone-induced postprandial carbohydrate intolerance in humans. Am J Physiol 1991;260:E513-E520.
73. Bak JF, Moller N, Schmitz O. Effects of growth hormone on fuel utilization and muscle glycogen synthase activity in normal humans. Am J Physiol 1991;260:E736-E742.
74. Press M. Growth hormone and metabolism. Diabetes Metab Rev 1988;4:391-414.
75. Dietz J, Schwartz J. Growth hormone alters lipolysis and hormone-sensitive lipase activity in 3T3-F442A adipocytes. Metabolism 1991;40:800-806.
76. Moller N, Schmitz O, Porksen N, Moller J, Jorgensen JO. Dose-response studies on the metabolic effects of a growth hormone pulse in humans. Metabolism 1992;41:172-175.
77. Mjaaland M, Unneberg K, Bjoro T, Revhaug A. Growth hormone treatment after abdominal surgery decreased carbohydrate oxidation and increased fat oxidation in patients with total parenteral nutrition. Metabolism 1993;42:185-190.
78. Eckel RH. Lipoprotein lipase. A multifunctional enzyme relevant to common metabolic diseases. N Engl J Med 1989;320:1060-1068.

79. Levak-Frank S, Radner H, Walsh A, et al. Muscle-specific overexpression of lipoprotein lipase causes a severe myopathy characterized by proliferation of mitochondria and peroxisomes in transgenic mice. J Clin Invest 1995;96:976-986.
80. Richelsen B, Pedersen SB, Moller-Pedersen T, Schmitz O, Moller N, Borglum JD. Lipoprotein lipase activity in muscle tissue influenced by fatness, fat distribution and insulin in obese females. Eur J Clin Invest 1993;23:226-233.
81. Oscarsson J, Ottosson M, Vikman-Adolfsson K, et al. GH but not IGF-I or insulin increases lipoprotein lipase activity in muscle tissues of hypophysectomised rats. J Endocrinol 1999;160:247-255.
82. Richelsen B, Kristensen K, Pedersen SB, Jorgensen JOL. Effects of GH treatment for 4 months in growth hormone deficient adults on adipose and muscle lipoprotein lipase activity and body composition. Int J Obesity 1997;21 Suppl 2:S39.
83. Froesch ER, Zapf J. Insulin-like growth factors and insulin: comparative aspects. Diabetologia 1985;28:485-493.
84. Czech MP. Structural and functional homologies in the receptors for insulin and the insulin-like growth factors. Cell 1982;31:8-10.
85. Ullrich A, Gray A, Tam AW, et al. Insulin-like growth factor I receptor primary structure: comparison with insulin receptor suggests structural determinants that define functional specificity. EMBO J 1986;5:2503-2512.
86. Caro JF, Poulos J, Ittoop O, Pories WJ, Flickinger EG, Sinha MK. Insulin-like growth factor I binding in hepatocytes from human liver, human hepatoma, and normal, regenerating, and fetal rat liver. J Clin Invest 1988;81:976-981.
87. Zapf J, Schoenle E, Waldvogel M, Sand I, Froesch ER. Effect of trypsin treatment of rat adipocytes on biological effects and binding of insulin and insulin-like growth factors: further evidence for the action of insulin-like growth factors through the insulin receptor. Eur J Biochem 1981;113:605-609.
88. Poggi C, Le Marchand-Brustel Y, Zapf J, Froesch ER, Freychet P. Effects and binding of insulin-like growth factor I in the isolated soleus muscle of lean and obese mice: comparison with insulin. Endocrinology 1979;105:723-730.
89. Lund S, Flyvbjerg A, Holman GD, Larsen FS, Pedersen O, Schmitz O. Comparative effects of IGF-I and insulin on the glucose transporter system in rat muscle. Am J Physiol 1994;267:E461-E466.
90. Valentini L, Holzenbein T, Winkler S, et al. Acute effects of insulin-like growth factor I on interorgan glucose and lactate flux in protein-catabolic dogs. J Surg Res 1995;59:606-613.
91. Lang CH. IGF-I stimulates muscle glucose uptake during sepsis. Shock 1996;5:22-27.
92. Elahi D, McAloon-Dyke M, Fukagawa NK, et al. Effects of recombinant human IGF-I on glucose and leucine kinetics in men. Am J Physiol 1993;265:E831-E838.
93. Leahy JL, Vandekerkhove KM. Insulin-like growth factor-I at physiological concentrations is a potent inhibitor of insulin secretion. Endocrinology 1990;126:1593-1598.
94. Guler HP, Zapf J, Froesch ER. Short-term metabolic effects of recombinant human insulin-like growth factor I in healthy adults. N Engl J Med 1987;317:137-140.
95. Baxter RC. Circulating binding proteins for the insulin like growth factors. Trends Endocrinol Metab 1993;4:91-96.
96. Ross R, Miell J, Freeman E, et al. Critically ill patients have high basal growth hormone levels with attenuated oscillatory activity associated with low levels of insulin-like growth factor-I. Clin Endocrinol (Oxf) 1991;35:47-54.

97. Hawker FH, Stewart PM, Baxter RC, et al. Relationship of somatomedin-C/insulin-like growth factor I levels to conventional nutritional indices in critically ill patients. Crit Care Med 1987;15:732-736.
98. Frayn KN, Price DA, Maycock PF, Carroll SM. Plasma somatomedin activity after injury in man and its relationship to other hormonal and metabolic changes. Clin Endocrinol (Oxf) 1984;20:179-187.
99. Fleming RY, Rutan RL, Jahoor F, Barrow RE, Wolfe RR, Herndon DN. Effect of recombinant human growth hormone on catabolic hormones and free fatty acids following thermal injury. J Trauma 1992;32:698-702.
100. Ross RJ, Buchanan CR. Growth hormone secretion: its regulation and the influence of nutritional factors. Nutr Res Rev 1990;3:143-162.
101. Spurrell EL, Yarwood GD, Coakley JH. IGF-I mRNA in skeletal muscle from critically patients. J Endocrinol 1996;148:P91.
102. Degerblad M, Almkvist O, Grunditz R, et al. Physical and psychological capabilities during substitution therapy with recombinant growth hormone in adults with growth hormone deficiency. Acta Endocrinol (Copenh) 1990;123:185-193.
103. Arafah BM. Reversible hypopituitarism in patients with large nonfunctioning pituitary adenomas. J Clin Endocrinol Metab 1986;62:1173-1179.
104. Vance ML. Hypopituitarism. N Engl J Med 1994;330:1651-1662.
105. Tanner JM, Whitehouse RH. The effect of human growth hormone on subcutaneous fat thickness in hyposomatotrophic and panhypopituitary dwarfs. J Endocrinol 1967;39:263-275.
106. Collipp PJ, Thomas J, Curti V, Sharma RK, Maddaiah VT, Cohn SE. Body composition changes in children receiving human growth hormone. Metabolism 1973;22:589-595.
107. Tanner JM, Hughes PC, Whitehouse RH. Comparative rapidity of response of height, limb muscle and limb fat to treatment with human growth hormone in patients with and without growth hormone deficiency. Acta Endocrinol (Copenh) 1977;84:681-696.
108. Laron Z. Deficiencies of growth hormone and somatomedins in man. In Cohen MP, Foa PP (Eds): Spec Top Endocrinol Metab 1983;5:149-199.
109. Van der Werff Ten Bosch JJ, Bot A. Effects of human pituitary growth hormone on body composition. Neth J Med 1987;30:220-227.
110. Parra A, Argote RM, Garcia G, Cervantes C, Alatorre S, Perez-Pasten E. Body composition in hypopituitary dwarfs before and during human growth hormone therapy. Metabolism 1979;1979 28:851-857.
111. Novak LP, Hayles AB, Cloutier MD. Effect of HGH on body composition of hypopituitary dwarfs. Four-compartment analysis and composite body density. Mayo Clin Proc 1972;47:241-246.
112. Tanner JM, Whitehouse RH, Hughes PC, Carter BS. Relative importance of growth hormone and sex steroids for the growth at puberty of trunk length, limb length, and muscle width in growth hormone-deficient children. J Pediatr 1976;89:1000-1008.
113. Rutherford OM, Jones DA, Round JM, Buchanan CR, Preece MA. Changes in skeletal muscle and body composition after discontinuation of growth hormone treatment in growth hormone deficient young adults. Clin Endocrinol (Oxf) 1991;34:469-475.
114. Brat O, Ziv I, Klinger B, Avraham M, Laron Z. Muscle force and endurance in untreated and human growth hormone or insulin-like growth factor-I-treated patients with growth hormone deficiency or Laron syndrome. Horm Res 1997;47:45-48.
115. Laron Z. Laron syndrome-From description to therapy. Endocrinologist 1993;3:21-28.

116.Leger J, Carel C, Legrand I, Paulsen A, Hassan M, Czernichow P. Magnetic resonance imaging evaluation of adipose tissue and muscle tissue mass in children with growth hormone (GH) deficiency, Turner's syndrome, and intrauterine growth retardation during the first year of treatment with GH. J Clin Endocrinol Metab 1994;78:904-909.
117.Leger J, Garel C, Fjellestad-Paulsen A, Hassan M, Czernichow P. Human growth hormone treatment of short-stature children born small for gestational age: effect on muscle and adipose tissue mass during a 3-year treatment period and after 1 year's withdrawal. J Clin Endocrinol Metab 1998;83:3512-3516.
118.Jorgensen JO, Pedersen SA, Thuesen L, et al. Beneficial effects of growth hormone treatment in GH-deficient adults. Lancet 1989;1:1221-1225.
119.Cuneo RC, Salomon F, Wiles CM, Sonksen PH. Skeletal muscle performance in adults with growth hormone deficiency. Horm Res 1990;33(4):55-60.
120.Jorgensen JO, Pedersen SA, Thuesen L, et al. Long-term growth hormone treatment in growth hormone deficient adults. Acta Endocrinol (Copenh) 1991;125:449-453.
121.Cuneo RC, Salomon F, Wiles CM, Hesp R, Sonksen PH. Growth hormone treatment in growth hormone-deficient adults. I. Effects on muscle mass and strength. J Appl Physiol 1991;70:688-694.
122.Binnerts A, Deurenberg P, Swart GR, Wilson JH, Lamberts SW. Body composition in growth hormone-deficient adults. Am J Clin Nutr 1992;55:918-923.
123.Rutherford OM, Beshyah SA, Johnston DG. Quadriceps strength before and after growth hormone replacement in hypopituitary adults: relationship to changes in lean body mass and IGF-I. Endocrinol Metab 1994;1:41-47.
124.Sartorio A, Narici M, Conti A, Monzani M, Faglia G. Quadriceps and hand-grip strength in adults with childhood-onset growth hormone deficiency. Eur J Endocrinol 1995;132:37-41.
125.Rutherford OM, Beshyah SA, Schott J, Watkins Y, Johnston DG. Histology of skeletal muscle in adults with GH deficiency: comparison with normal muscle and response to GH treatment. Clin Sci (Colch) 1995;88:67-71.
126.Janssen YJ, Doornbos J, Roelfsema F. Changes in muscle volume, strength, and bioenergetics during recombinant human growthhormone (GH) therapy in adults with GH deficiency. J Clin Endocrinol Metab 1999;84:279-284.
127.Bottinelli R, Narici M, Pellegrino MA, et al. Contractile properties and fiber type distribution of quadriceps muscles in adults with childhood-onset growth hormone deficiency. J Clin Endocrinol Metab 1997;82:4133-4138.
128.Whitehead HM, Gilliland JS, Allen IV, Hadden DR. Growth hormone treatment in adults with growth hormone deficiency: effect on muscle fibre size and proportions. Acta Paediatr Scand 1989;356:65-67.
129.Cuneo RC, Salomon F, Wiles CM, et al. Contractile properties of the quadriceps muscle in growth hormone-deficient hypopituitary adults. Horm Res 1992;37:23-28.
130.Pette D. Metabolic heterogeneity of muscle fibres. J Exp Biol 1985;115:179-189.
131.Marie P. Sur deux cas d'acromegalie; hypertrophie singuliere, non congenitale, des extremites superieures, inferieures et cephalique. Rev de Med (Paris) 1886;6:297-333.
132.Souza-Leite JD. Essays on acromegaly. London, New Sydenham Society 1891:1-28.
133.Souza-Leite JD 1891 A thesis on acromegaly (Marie's malady). pp 31-160 In Marie P, Souza-Leite JD: Essays on acromegaly. London, New Sydenham Society pp 31-160.
134.Mastaglia FL, Barwich DD, Hall R. Myopathy in acromegaly. Lancet 1970;2:907-909.

135.Lundberg PO, Osterman PO, Stalberg E. Neuromuscular signs and symptoms in acromegaly. In Walton JN, Canal N, Scarlata G et al (eds): International Congress on Muscular Disease. Milan, Excerpta Medica Fundation 1970:531-534.
136.Pickett JB, Layzer RB, Levin SR, Scheider V, Campbell MJ, Sumner AJ. Neuromuscular complications of acromegaly. Neurology 1975;25:638-645.
137.Greenbaum AL, Young FG. Distribution of protein in tissues of rats treated with anterior-pituitary growth-hormone. Nature (Lond) 1950;165:521-522.
138.Wolf E, Wanke R, Schenck E, Hermanns W, Brem G. Effects of growth hormone overproduction on grip strength of transgenic mice. Eur J Endocrinol 1995;133:735-740.
139.Costa C, Solanes G, Visa J, Bosch F. Transgenic rabbits overexpressing growth hormone develop acromegaly and diabetes mellitus. FASEB J 1998;12:1455-1460.
140.Prysor-Jones RA, Jenkins JS. Effect of excessive secretion of growth hormone on tissues of the rat, with particular reference to the heart and skeletal muscle. J Endocrinol 1980;85:75-82.
141.Ayling CM, Moreland BH, Zanelli JM, Schulster D. Human growth hormone treatment of hypophysectomized rats increases the proportion of type-1 fibres in skeletal muscle. J Endocrinol 1989;123:429-435.
142.Ayling CM, Zanelli JM, Moreland BH, Schulster D. Effect of human growth hormone injection on fibre type composition and matabolic activity in a skeletal muscle from normal and hypophysectomized rats. Growth Regul 1992;2:133-143.
143.Stern LZ, Payne CM, Hannapel LK. Acromegaly: histochemical and electron microscopic changes in deltoid and intercostal muscle. Neurology 1974;24:589-593.
144.Mastaglia FL. Pathological changes in skeletal muscle in acromegaly. Acta Neuropathol (Berl) 1973;24:273-286.
145.Nagulesparen M, Trickey R, Davies MJ, Jenkins JS. Muscle changes in acromegaly. Br Med J 1976;2:914-915.
146.Christiansen JS, Jorgensen JO. Beneficial effects of GH replacement therapy in adults. Acta Endocrinol (Copenh) 1991;125:7-13.
147.Cuneo RC, Salomon F, McGauley GA, Sonksen PH. The growth hormone deficiency syndrome in adults. Clin Endocrinol (Oxf) 1992;37:387-397.
148.Attanasio AF, Lamberts SW, Matranga AM, et al. Adult growth hormone (GH)-deficient patients demonstrate heterogeneity between childhood onset and adult onset before and during human GH treatment. Adult Growth Hormone Deficiency Study Group. J Clin Endocrinol Metab 1997;82:82-88.
149.Bengtsson BA, Eden S, Lonn L, et al. Treatment of adults with growth hormone (GH) deficiency with recombinant human GH. J Clin Endocrinol Metab 1993;76:309-317.
150.Whitehead HM, Boreham C, McIlrath EM, et al. Growth hormone treatment of adults with growth hormone deficiency: results of a 13-month placebo controlled cross-over study. Clin Endocrinol (Oxf) 1992;36:45-52.
151.Binnerts A, Swart GR, Wilson JH, et al. The effect of growth hormone administration in growth hormone deficient adults on bone, protein, carbohydrate and lipid homeostasis, as well as on body composition. Clin Endocrinol (Oxf) 1992;37:79-87.
152.Beshyah SA, Freemantle C, Shahi M, et al. Replacement treatment with biosynthetic human growth hormone in growth hormone-deficient hypopituitary adults. Clin Endocrinol (Oxf) 1995;42:73-84.
153.Jorgensen JO, Vahl N, Hansen TB, Thuesen L, Hagen C, Christiansen JS. Growth hormone versus placebo treatment for one year in growth hormone deficient adults:

increase in exercise capacity and normalization of body composition. Clin Endocrinol (Oxf) 1996;45:681-688.
154.ter Maaten JC, de Boer H, Kamp O, Stuurman L, Van Der Veen EA. Long-term effects of growth hormone (GH) replacement in men with childhood-onset GH deficiency. J Clin Endocrinol Metab 1999;84:2373-2380.
155.Gibney J, Wallace JD, Spinks T, et al. The effects of 10 years of recombinant human growth hormone (GH) in adult GH-deficient patients. J Clin Endocrinol Metab 1999;84:2596-2602.
156.Moller J, Jorgensen JO, Moller N, Stodkilde-Jorgensen H, Christiansen JS. Assessment of fat and muscle volume and water content by magnetic resonance imaging (MRI) and photon spectroscopy in growth hormone (GH) deficient adults before and after GH treatment. Endocrinology and Metabolism 1994;1:131-135.
157.Orme SM, Sebastian JP, Oldroyd B, et al. Comparison of measures of body composition in a trial of low dose growth hormone replacement therapy. Clin Endocrinol (Oxf) 1992;37453-459.
158.Jorgensen JO, Thuesen L, Muller J, Ovesen P, Skakkebaek NE, Christiansen JS. Three years of growth hormone treatment in growth hormone-deficient adults: near normalization of body composition and physical performance. Eur J Endocrinol 1994;130:224-228.
159.Johannsson G, Grimby G, Sunnerhagen KS, Bengtsson BA. Two years of growth hormone (GH) treatment increase isometric and isokinetic muscle strength in GH-deficient adults. J Clin Endocrinol Metab 1997;82:2877-2884.
160.Wallymahmed ME, Foy P, Shaw D, Hutcheon R, Edwards RH, MacFarlane IA. Quality of life, body composition and muscle strength in adult growth hormone deficiency: the influence of growth hormone replacement therapy for up to 3 years. Clin Endocrinol (Oxf) 1997;47:439-446.
161.Goldberg AL. Work-induced growth of skeletal muscle in normal and hypophysectomized rats. Am J Physiol 1967;213:1193-1198.
162.Zanconato S, Moromisato DY, Moromisato MY, et al. Effect of training and growth hormone suppression on insulin-like growth factor I mRNA in young rats. J Appl Physiol 1994;76:2204-2209.
163.Pell JM, Bates PC. Collagen and non-collagen protein turnover in skeletal muscle of growth hormone-treated lambs. J Endocrinol 1987;115:R1-R4.
164.Caidahl K, Eden S, Bengtsson BA. Cardiovascular and renal effects of growth hormone. Clin Endocrinol (Oxf) 1994;40:393-400.
165.Narici MV, Roi GS, Landoni L, Minetti AE, Cerretelli P. Changes in force, cross-sectional area and neural activation during strength training and detraining of the human quadriceps. Eur J Appl Physiol 1989;59:310-319.
166.Davies CT, Thomas DO, White MJ. Mechanical properties of young and elderly human muscle. Acta Med Scand Suppl 1986;711:219-226.
167.Lindle RS, Metter EJ, Lynch NA, et al. Age and gender comparisons of muscle strength in 654 women and men aged 20-93 yr. J Appl Physiol 1997;83:1581-1587.
168.Jubrias SA, Odderson IR, Esselman PC, Conley KE. Decline in isokinetic force with age: muscle cross-sectional area and specific force. Pflugers Arch 1997;434:246-253.
169.Larsson L, Grimby G, Karlsson J. Muscle strength and speed of movement in relation to age and muscle morphology. J Appl Physiol 1979;46451-456.
170.Cupido CM, Hicks AL, Martin J. Neuromuscular fatigue during repetitive stimulation in elderly and young adults. Eur J Appl Physiol 1992;65:567-572.

171.Doherty TJ, Brown WF. Age-related changes in the twitch contractile properties of human thenar motor units. J Appl Physiol 1997;82:93-101.
172.Doherty TJ, Vandervoort AA, Taylor AW, Brown WF. Effects of motor unit losses on strength in older men and women. J Appl Physiol 1993;74:868-874.
173.Kallman DA, Plato CC, Tobin JD. The role of muscle loss in the age-related decline of grip strength: cross-sectional and longitudinal perspectives. J Gerontol 1990;45:M82-88.
174.Reed RL, Pearlmutter L, Yochum K, Meredith KE, Mooradian AD. The relationship between muscle mass and muscle strength in the elderly. J Am Geriatr Soc 1991;39:555-561.
175.Evans WJ. Effects of exercise on body composition and functional capacity of the elderly. J Gerontol A Biol Sci Med Sci 1995;50:147-50.
176.Fleg JL, Lakatta EG. Role of muscle loss in the age-associated reduction in VO2 max. J Appl Physiol 1988;65:1147-1151.
177.Flynn MA, Nolph GB, Baker AS, Krause G. Aging in humans: a continuous 20-year study of physiologic and dietary parameters. J Am Coll Nutr 1992;11:660-672.
178.Lexell J, Henriksson-Larsen K, Winblad B, Sjostrom M. Distribution of different fiber types in human skeletal muscles: effects of aging studied in whole muscle cross sections. Muscle Nerve 1983;6:588-595.
179.Larsson L. Histochemical characteristics of human skeletal muscle during aging. Acta Physiol Scand 1983;117:469-471.
180.Welle S, Thornton C, Jozefowicz R, Statt M. Myofibrillar protein synthesis in young and old men. Am J Physiol 1993;264:E693-E698.
181.Rantanen T, Avela J. Leg extension power and walking speed in very old people living independently. J Gerontol A Biol Sci Med Sci 1997;52:M225-M231.
182.Corpas E, Harman SM, Blackman MR. Human growth hormone and human aging. Endocr Rev 1993;14:20-39.
183.Rudman D, Kutner MH, Rogers CM, Lubin MF, Fleming GA, Bain RP. Impaired growth hormone secretion in the adult population: relation to age and adiposity. J Clin Invest 1981;67:1361-1369.
184.Holloway L, Butterfield G, Hintz RL, Gesundheit N, Marcus R. Effects of recombinant human growth hormone on metabolic indices, body composition, and bone turnover in healthy elderly women. J Clin Endocrinol Metab 1994;9:470-479.
185.Toogood AA, O'Neill PA, Shalet SM. Beyond the somatopause: growth hormone deficiency in adults over the age of 60 years. J Clin Endocrinol Metab 1996;81:460-465.
186.Sonntag WE, Hylka VW, Meites J. Growth hormone restores protein synthesisi in skeletal muscle of old male rats Endocrinology 1984;11075:1875-1879.
187.Marcus R, Butterfield G, Holloway L, et al. Effects of short term administration of recombinant human growth hormone to elderly people. J Clin Endocrinol Metab 1990;70:519-527.
188.Rudman D, Feller AG, Nagraj HS, et al. Effects of human growth hormone in men over 60 years old. N Engl J Med 1990;323:1-6.
189.Welle S, Thornton C, Statt M, McHenry B. Growth hormone increases muscle mass and strength but does not rejuvenate myofibrillar protein synthesis in healthy subjects over 60 years old. J Clin Endocrinol Metab 1996;81:3239-3243.

190.Butterfield GE, Thompson J, Rennie MJ, Marcus R, Hintz RL, Hoffman AR. Effect of rhGH and rhIGF-I treatment on protein utilization in elderly women. Am J Physiol 1997;272:E94-E99.
191.Yarasheski KE, Zachwieja JJ, Campbell JA, Bier DM. Effect of growth hormone and resistance exercise on muscle growth and strength in older men. Am J Physiol 1995;268:E268-E276.
192.Taaffe DR, Pruitt L, Reim J, et al. Effect of recombinant human growth hormone on the muscle strength response to resistance exercise in elderly men. J Clin Endocrinol Metab 1994;79:1361-1366.
193.Taaffe DR, Jin IH, Vu TH, Hoffman AR, Marcus R. Lack of effect of recombinant human growth hormone (GH) on muscle morphology and GH-insulin-like growth factor expression in resistance-trained elderly men. J Clin Endocrinol Metab 1996;81:421-425.
194.Papadakis MA, Grady D, Black D, et al. Growth hormone replacement in healthy older men improves body composition but not functional ability. Ann Intern Med 1996;124:708.
195.Lieberman SA, Hoffman AR. Sequelae to acromegaly: reversibility with treatment of the primary disease. Horm Metab Res 1990;22:313-318.
196.Von Werder K. Growth hormone therapy in prostatic cancer:preliminary results. In Ranke MB, Christiansen JS (eds), The complexity of endocrine systems. J&J Verlag, Mannheim 1996.
197.Murphy LJ, Bell GI, Friesen HG. Growth hormone stimulates sequential induction of c-myc and insulin-like growth factor I expression in vivo. Endocrinology 1987;120:1806-1812.
198.Jenkins P. Cancer in acromegaly Trends Endocrinol Metab 1998 9:360-366.
199.Chan JM, Stampfer MJ, Giovannucci E, et al. Plasma insulin-like growth factor-I and prostate cancer risk: a prospective study. Science 1998;279:563-566.

15

GROWTH HORMONE AND NUTRITIONAL STATUS

Massimo Scacchi, Angela I. Pincelli, Francesco Cavagnini
Chair of Endocrinology, University of Milan, IRCCS Ospedale S Luca, Istituto Auxologico Italiano, Milan, Italy

INTRODUCTION

Growth hormone (GH) secretion is controlled by central and peripheral factors (1). The chief hypothalamic regulators of GH release are growth hormone releasing hormone (GHRH) and somatostatin (somatotropin release inhibiting hormone, SRIH). The finding of specific hypothalamic and pituitary receptors for the recently synthesized GH secretagogues (GHS) has led to the hypothesis of a third neuroendocrine pathway affecting GH secretion mediated by a still unidentified endogenous ligand for GHS receptors. A number of other neuroendocrine signals represented by neurotransmitters, neuropeptides and neurohormones may also intervene on the above hypophysiotropic hormones indirectly modulating GH secretion. Among the peripheral factors a pivotal role is played by the negative feedback exerted at both hypothalamic and pituitary levels by insulin-like growth factor I (IGF-I), the chief mediator of most biological actions of GH. GH is endowed with a relevant role in the regulation of body composition and energy expenditure, and thus it is not surprising that metabolic substrates (glucose, aminoacids and lipids) exert a feedback control on GH release. This chapter will mainly focus on the functional status of the GH-IGF-I axis in conditions of under- and overnutrition. In this context the effects of metabolic substrates on GH secretion as well as the influence exerted by GH on body composition and energy metabolism will be discussed.

ROLE OF MACRONUTRIENTS IN THE CONTROL OF GH RELEASE

Glucose is a major regulator of GH secretion. In particular, in humans a decrease in blood glucose is followed by a significant rise in GH serum levels, whereas hyperglycaemia inhibits GH release. The observations that GHRH pretreatment prevents the somatotropin response to a subsequent GHRH bolus but not to insulin-induced hypoglycaemia and that GHRH administration and hypoglycaemia have an additive effect in releasing GH suggest for hypoglycaemia a mechanism of action mediated by the reduction of SRIH release. Opposite to hypoglycemia, increased blood glucose levels inhibit both basal and GHRH-stimulated GH secretion likely via central SRIH release. In line with these concepts, administration of pyridostigmine, an indirect cholinergic agonist thought to inhibit hypothalamic SRIH secretion, counteracts the inhibitory effect of glucose on GH release. In contrast to what observed in humans, in rats hyperglycaemia does not affect basal or GHRH-induced GH release while the stress of hypoglycaemia reduces GH secretion. These findings disclose major differences between the two species in the responsiveness of GH to blood glucose variations. Similarly, discrepancies exist between humans and rodents regarding the somatotropin reactivity to calorie restriction (see below). Among the peripheral factors contributing to the regulation of GH release, free fatty acids (FFA) play an important role by exerting a negative feedback at both the hypothalamus, likely via enhancement of SRIH tone, and the pituitary. This latter action is well documented by in vitro studies showing an inhibitory effect of FFA on somatotropin release from cultured rat somatotropes (2). Likewise, in humans an increase in FFA serum levels blunts the GH responses to both GHRH and suprapituitary stimuli. Aminoacids also participate in the control of GH secretion. High protein meals as well as administration of basic (arginine and ornithine) and aromatic (tryptophan) aminoacids enhance GH release in normal subjects. The observations that arginine infusion potentiates the somatotropin response to a maximal dose of GHRH but not to pyridostigmine and that pretreatment with GHRH does not prevent the GH response to subsequent arginine application all point to a hypothalamic site of action for this aminoacid, likely through inhibition of SRIH release. Contrary to what reported in humans, in rats aminoacids do not affect GH secretion.

INFLUENCE OF GH ON BODY COMPOSITION AND ENERGY METABOLISM

Through its metabolic actions GH greatly influences body composition. The lipolytic effect of somatotropin has long been known: its administration to animals and humans increases serum concentrations of glycerol and FFA. Furthermore, experimental data suggest a stimulatory influence of GH on fat cell formation in clonal preadipocyte cell lines, thus amplifying GH-induced lipolysis. The protein anabolic properties of somatotropin are also well documented: indeed, GH stimulates aminoacid uptake and protein synthesis. This anabolic action is testified by the nitrogen retention taking place in subjects receiving the hormone. Among the effects of GH, at least other two are important for body composition: its antinatriuretic action leading to sodium and water retention, and the stimulation of bone turnover with bone formation prevailing on resorption. The lack or excess of these GH actions are reflected by the changes of body composition observed in clinical models of somatotropin hypo- or hypersecretion. Adult patients with GH deficiency display increased abdominal adipose tissue, now regarded as a risk factor for cardiovascular disease, reduced muscle mass, dehydration due to decreased body extracellular water and osteopenia with increased risk for fracture (3). In these patients replacement therapy with recombinant human GH (rhGH) is followed by improvement or normalization of body composition. Conversely, in acromegaly a decrease in body fat, associated with an increase in lean mass and extracellular water have been reported (4). Once again, these abnormalities are reversed by treatment.

LEPTIN AND GH SECRETION

Leptin, the product of the adipocyte specific *ob* gene, is thought to exert a pivotal role in the control of body weight, signalling to the hypothalamus the amount of fat in the body, i.e. the nutritional status. This function is mainly carried out via a decrease in food consumption, likely mediated by inhibition of neuropeptide Y (NPY) secretion, and an enhancement of energy expenditure. In humans, leptin levels correlate with the percentage of body fat and with the body mass index. Thus, states of fat accumulation such as obesity, Cushing's syndrome and adult GH deficiency are associated with elevated serum leptin concentrations (5-7); conversely, circulating leptin is low in states of malnutrition such as anorexia nervosa

(8). A huge body of evidence has accumulated showing that in rodents, but only exceptionally in humans (9,10), massive obesity may develop as a result of mutations of the leptin gene, as in ob/ob mice (11), or receptor, as in the db/db mice (12). Considering the influence exerted by leptin and GH on body composition, studies on the functional relationships between the two hormones have been undertaken. Intracerebroventricular (i.c.v.) injection of leptin antiserum to fed rats decreases spontaneous GH release, whereas i.c.v. administration of leptin to fasted rats has the opposite effect (13). These actions appear to be exerted at the hypothalamus through modifications of GHRH and SRIH release. In fact, administration of GHRH or SRIH antiserum abolishes and enhances, respectively, the leptin-induced GH secretion. Furthermore, in vitro, leptin induces a time dependent decrease in basal SRIH release from rat hypothalamic neurons. In all, the available observations point to leptin as an at least permissive factor for GH release in rodents. This effect is likely mediated by a decreased secretion of central NPY, a known inhibitor of GH release. Accordingly, i.c.v. injection of NPY has been shown to prevent leptin-induced GH release in fasted rats. No convincing evidence that GH may influence leptin secretion has been obtained so far. In vitro, neither GH nor IGF-I are able to affect leptin release from isolated rat adipocytes (14). Furthermore, the reduction of basally elevated serum leptin observed in GH-deficient adults after treatment with rhGH is likely secondary to the fat mass-reducing action of the hormone (7).

NUTRITIONAL STATUS AND GH SECRETION. ANIMAL MODELS

In most mammals including man, fasting or caloric restriction enhances basal GH secretion. An exception is represented by the rat, in which prolonged fasting decreases serum GH levels. Several mechanisms likely contribute to the somatotropin hypersecretion induced by fasting and calorie restriction: among them, reduced IGF-I production with consequent lack of negative feedback on GH release, and changes in hypothalamic output of GHRH and SRIH. Food restriction appears to increase GH secretion in sheep by reducing SRIH release into hypophysial portal blood. Furthermore, data obtained in calorie restricted dogs favour the hypothesis of a reduced SRIH tone. In fact, pyridostigmine, an indirect cholinergic agonist thought to inhibit SRIH release, is able to potentiate the somatotropin response to GHRH in fed but not in fasted dogs. An alternative explanation for this finding is a possible fasting-induced increase in hypothalamic cholinergic tone which would render pyridostigmine unable to further enhance cholinergic neurotransmission. The involvement of SRIH is also suggested by the observation that pharmacological manipulations of the cholinergic tone fail to affect the GH response to GHRH in conscious male rats with anterolateral deafferentation of the mediobasal hypothalamus or treated with cysteamine, the two experimental models of hypothalamic SRIH depletion. In opposition to what described in other mammals, prolonged food deprivation inhibits spontaneous GH secretion in rats. Immunocytochemical and molecular analysis disclosed a general down-regulation of the hypothalamic-pituitary axis for GH secretion in food-deprived adult rats with decreased GHRH and SRIH immunostaining in the median eminence of the hypothalamus and reduced levels of mRNA for GHRH and SRIH in the hypothalamus and for GH in the anterior pituitary. The finding that central administration of an antiserum directed against NPY, a peptide involved in the control of both food intake (stimulation) and GH secretion (inhibition), improves somatotropin pulsatility in food-deprived adult rats suggests for NPY a possible role in fasting-induced GH hyposecretion (15). Refeeding after prolonged food restriction restores spontaneous GH secretion in rats. In particular, the amount of dietary protein appears to be essential for the reinstatement of the ultradian rhythmicity, a concept supported by results obtained with parenteral administration of aminoacids (15). Refeeding is also followed by normalization of IGF-I levels, which are reduced in calorie restricted rats likely due to both hyposomatotropism and decreased GH binding or post-receptor defects in the liver (16). At variance with spontaneous GH secretion, the hormonal response to GHRH is enhanced in fasted rats. This might be due to increased sensitivity to the releasing hormone, since both GHRH receptor mRNA and GHRH binding are increased in the pituitary of food restricted adult rats (17). The influence of overweight on GH secretion has been extensively investigated in the genetically obese Zucker rat. As in humans, spontaneous GH release is markedly impaired in these animals compared to their lean siblings. Furthermore, in Zucker rats a blunted somatotropin response to GHRH has been reported in vitro, together with reduced hypothalamic GHRH and, as a likely consequence, pituitary GH content and expression (18,19). This profound alteration of the hypothalamic-somatotropic axis is like to be linked to precocious resistance to leptin in genetically obese rats.

NUTRITIONAL STATUS AND GH SECRETION. HUMAN MODELS

Starvation and fasting

Chronic undernutrition (marasmus and kwashiorkor) and acute dietary restriction (fasting) induce a marked reduction of plasma IGF-I levels in humans. In children with severe protein restriction (kwashiorkor) or protein-energy deprivation (marasmus) growth retardation due to lack of IGF-I is a common feature. Low IGF-I levels have also been reported in other forms of malnourishment such as severe inflammatory bowel disease, celiac disease and anorexia nervosa. In general, the degree of IGF-I reduction relates to the severity of the nutritional insult, and IGF-I levels significantly increase with nutritional rehabilitation. Both energy and protein content of the diet are important for restoration of plasma IGF-I concentrations to normal after caloric restriction. Indeed, an aminoacid rich diet accelerates the return of IGF-I to normal and improves nitrogen balance. In chronic malnutrition, the low IGF-I plasma concentrations in spite of normal or elevated GH serum levels envisage a condition of peripheral resistance to GH. In normal humans fasting induces a clear-cut decrease in IGF-I levels. These latter fall to 15-20% of baseline values after 10 days of fast, and return to normal levels with refeeding (20). Fasted humans also display an impaired IGF-I response to exogenous GH administration (21). Studies employing computerized algorithms disclose an enhancement of spontaneous somatotropin secretion in fasted volunteers (22,23). In particular, the rise of GH production rate is accounted for by an increase in both the number of secretory bursts and the mass of hormone secreted per burst. In fasted humans the infusion of recombinant human IGF-I (rhIGF-I) suppresses the enhanced pulsatile GH secretion with a similar time course as refeeding, suggesting that changes in plasma IGF-I concentrations mediate the effects of nutrition on GH release (24). The influence of nutrition on IGF-I production is such a strong one that a 6-day fast has been shown to reduce plasma somatomedin to 65% of baseline even in acromegalic patients (25). Interestingly, the decrease in IGF-I induced by calorie restriction is much smaller in obese patients than in normal subjects (16,26). Thus, in obesity the production of IGF-I appears to be relatively independent from energy intake, provided that a sufficient protein supply is maintained.

Anorexia nervosa

This psychiatric disease represents a peculiar and extensively studied pathophysiological model of GH secretion in humans. Like in other malnourishment syndromes, in anorexia nervosa low IGF-I plasma levels are associated with high baseline GH concentrations (27). However, the somatotropin responsiveness to different provocative stimuli is highly variable in this clinical condition. The GH rise, often exaggerated after GHRH (28), is generally normal following clonidine (29) and arginine (30), and definitely blunted in response to insulin-induced hypoglycaemia (31), dopaminergic agonists (32) and acute glucocorticoid administration (33). Failure of glucose infusion and of pharmacological manipulation of the cholinergic tone (i.e. administration of the muscarinic receptor blocker pirenzepine or the cholinesterase inhibitor pyridostigmine) to affect the GH response to GHRH can also be observed. In addition, anorectic patients often display paradoxical GH responses to thyrotropin releasing hormone and gonadotropin releasing hormone, a behaviour shared with other conditions of tumoral and non tumoral hypersomatotropinism such as acromegaly, renal failure and diabetes mellitus. The variability of GH responsiveness to pharmacological challenges suggests a derangement in the hypothalamic neuroregulation of somatotropin release in anorexia nervosa. In particular, the lack of GH reactivity to factors known to affect central SRIH release (blood glucose changes, glucocorticoid administration) points to a defect in hypotalamic SRIH tone. This hypothesis is supported by the failure of pirenzepine and pyridostigmine to affect GHRH-induced GH release: indeed, such a behaviour is observed in rats with hypothalamic SRIH depletion (32). Additional indirect support to this view is provided by a recent study of ours on spontaneous GH secretion in anorexia nervosa. Using the pulse detection algorithm Cluster (34), we have shown that the nocturnal GH release of anorectic patients is enhanced (Fig.1) with an increase in peak frequency and interpeak hormonal concentrations (Table 1).

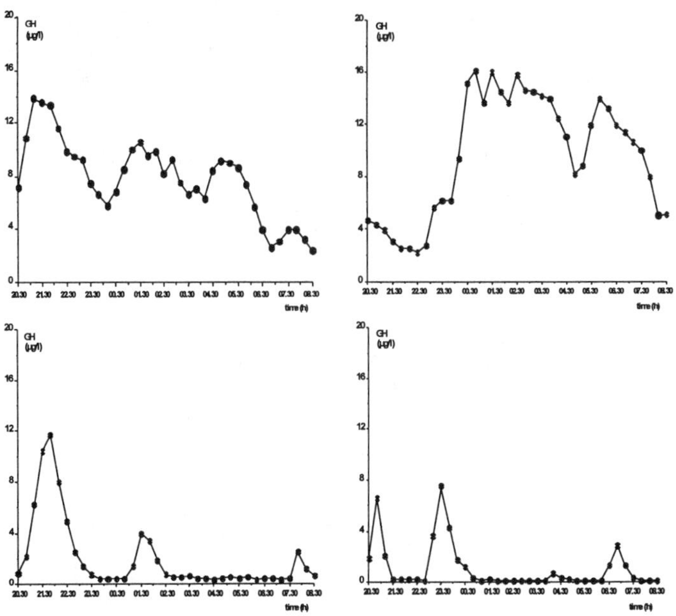

Figure 1 Growth hormone serum profiles in two representative anorectic patients (upper panel) and in two control women (lower panel).

Table 1 Mean GH pulse parameters determined by Cluster analysis in anorectic patients and normal subjects.

	Anorectic patients	Normal women	*p*
Peak frequency (number/12h)	5.0 ± 0.81	3.0 ± 0.89	< 0.01
Peak width (min)	108.2 ± 16.71	101.6 ± 36.84	NS
Peak height (µg/l)	8.6 ± 3.37	5.1 ± 2.00	< 0.05
Incremental peak height (µg/l)	3.0 ± 1.39	4.2 ± 1.71	NS
Interpeak interval (min)	114.1 ± 27.78	297.2 ± 121.08	< 0.01
Valley mean (µg/l)	5.9 ± 2.25	1.0 ± 1.30	< 0.01
Pulse area ($\mu g \cdot l^{-1} \cdot min^{-1}$)	323.9 ± 182.74	261.2 ± 149.16	NS
Total AUC ($\mu g \cdot l^{-1} \cdot min^{-1}$)	4743 ± 1520.09	1148.6 ± 519.27	< 0.01
Pulsatile AUC ($\mu g \cdot l^{-1} \cdot min^{-1}$)	1530.4 ± 654.72	769.8 ± 404.02	< 0.05
Basal AUC ($\mu g \cdot l^{-1} \cdot min^{-1}$)	3212.5 ± 990.45	378.7 ± 123.27	< 0.01
Pulsatile/total AUC	0.31 ± 0.06	0.62 ± 0.16	< 0.05

In our study, the enhanced GH release was mainly due to an increase in its nonpulsatile component, which represented up to 70% of somatotropin secretion in our patients. The pulsatile fraction of GH release was also increased, due to the higher number of secretory episodes and despite the normal amount of hormone released per burst. The possibility that the elevated interpeak GH levels of anorectic patients were due to a decrease in the hormonal metabolic clearance rate was subsequently ruled out by deconvolution analysis of the data (35), which demonstrated a comparable GH half-life in anorectic patients and normal women. Along the same line, the GH basal secretion rate was significantly higher in anorectic women than in controls. This abnormality, together with the confirmed increase in secretory burst frequency, accounted for a 3.5-fold higher GH production rate in patients compared to healthy women. In addition, anorectic patients displayed elevated values of approximate entropy (ApEn), testifying a disordered pattern of GH secretion (Table 2).

Table 2 Mean GH pulse parameters determined by deconvolution analysis in anorectic patients and normal subjects.

	Anorectic patients	Normal women	*p*
Number of secretory bursts (number/12h)	9.1 ± 0.46	5.0 ± 0.86	<0.01
Half-duration (min)	45.7 ± 6.2	33.6 ± 6.9	NS
Amplitude (µg/l)	0.47 ± 0.19	0.31 ± 0.07	NS
Mass (µg/l)	15.8 ± 2.2	10.2 ± 2.5	NS
Interbust interval (min)	69.0 ± 3.9	174 ± 34.7	<0.01
Half-life (min)	15.4 ± 1.6	15.5 ± 1.5	NS
Production rate ($\mu g \cdot l^{-1} \cdot min^{-1}$)	146 ± 22.9	41.3 ± 7.8	<0.01
Integrated AUC ($\mu g \cdot l^{-1} \cdot min^{-1}$)	4722 ± 570	1149 ± 212	<0.01
Basal secretion ($\mu g \cdot l^{-1} \cdot min^{-1}$)	0.133 ± 0.049	0.010 ± 0.002	<0.01
Mean concentration (µg/l)	6.6 ± 0.8	1.5 ± 0.29	<0.01
ApEn (1.20%)	0.824 ± 0.059	0.454 ± 0.042	<0.01

Similar results have been reported more recently by others (36). The increase in GH pulse frequency might be accounted for by an increased frequency of GHRH discharges, while the elevated interpulse GH levels are compatible with a reduction of hypothalamic SRIH tone. Indeed, low cerebrospinal fluid levels of SRIH have been reported in anorectic patients (37). However, the reduction of SRIH tone does not seem to be absolute, since a release of the neuropeptide is like to be elicited by adequate stimuli: this is suggested by the lack of GH response to repeated GHRH stimulation, which would indicate a substantial integrity of the SRIH-mediated GH autofeedback in anorexia nervosa. An altered GH responsiveness to the GHS hexarelin has been recently reported in anorexia nervosa (38). Personal preliminary data seem to confirm this abnormality: in our experience (35), at variance with what observed in normal subjects, the GH response to hexarelin in anorectic women was significantly smaller than the one elicited by GHRH; furthermore, the synthetic peptide failed to potentiate the GHRH-induced GH release in the same patients. Considering the possible mechanisms of action of GHS (anti-SRIH effect and/or binding to hypothalamic and pituitary receptors for an endogenous ligand), these findings could be interpreted as due to a defective SRIH tone and/or an enhanced secretion of the postulated endogenous ligand for GHS receptors. The latter hypothesis is supported by the demonstration of a concomitant impairment of the ACTH response to hexarelin in anorectic women (39). Altogether, these data suggest also a derangement of the third neuroendocrine pathway controlling GH release in this clinical condition. The disordered spontaneous GH secretion of anorectic patients, characterized by ApEn values similar to those observed in acromegaly, might be due to abnormalities in IGF-I feedback (36). Studies investigating this possibility are in progress in our laboratory. Abnormalities of spontaneous GH release similar to those disclosed by Cluster analysis in anorectic patients, i.e. increased peak frequency and interpeak GH levels with unchanged individual pulse area, have been reported in another condition of cellular fasting, i.e. in poorly controlled diabetes mellitus. Interestingly, the pattern of spontaneous GH secretion in anorexia appears to be quite different from that described in critically ill patients, who display low values of secretory burst amplitude and GH amount per secretory episode. This divergence may reflect pathophysiological differences, since anorexia nervosa, even when associated with important malnutrition and reduction of total body protein, is not characterized by the markedly increased protein catabolism, causing vital tissue wasting with fat depot preservation, typically observed in long lasting critical illness. However, recent studies seem to distinguish two different subpopulations of anorectic patients as regards the GH secretory pattern, namely hyper- and hyposecretors (26). Further studies are needed to gain new insights into these pathophysiological mechanisms. As for the peripheral limb of the GH-IGF-I axis in anorexia nervosa, the low IGF-I levels in spite of elevated somatotropin concentrations point to a peripheral resistance to GH action, in agreement with what reported in other malnourishment syndromes. This view is supported by personal data, showing no significant increase in IGF-I plasma levels in anorectic patients following subcutaneous administration of rhGH, 4U daily for four consecutive days (40). As it suits to a condition of acquired GH resistance, the circulating levels of the high affinity GH binding protein (GHBP), corresponding to the extracellular domain of the GH receptor, have been reported to be low in anorexia nervosa (27). Thus, the effect of malnutrition to decrease GH receptor number and/or function would explain many of the alterations of the GH-IGF-I axis observed in anorexia, i. e. decreased production of IGF-I and of the GH- and IGF-dependent IGF binding protein (IGFBP)-3 and increased GH secretion due to reduced negative feedback by IGF-I. As expected, IGFBP-1 and IGFBP-2, inversely correlated to serum insulin, are elevated in anorexia, a condition in which undernutrition leads to hypoinsulinemia (27). However, the lack of a negative correlation between IGF-I levels and the parameters of spontaneous GH secretion observed in our group of patients (34) merges with the absence of correlation between somatomedin concetrations and magnitude of GHRH-induced GH release reported by others (41), suggesting that factors other than or additional to the malnutrition-dependent lowering of IGF-I are involved in the altered GH secretion in anorexia nervosa. Along the same line is the absence of a negative correlation between parameters of spontaneous GH release and body mass index (BMI) in the same patients.

Obesity

Recent studies in normal subjects have established close relationships between parameters of spontaneous GH release and variables such as age, fitness and degree of adiposity. In particular, negative correlations were established between BMI on one side and GH half-life, amplitude of GH secretory episodes and pulsatile component of GH release on the other side. Furthermore, the percentage of body fat was negatively correlated with 24h GH integrated concentration, GH amount per secretory burst and GH half-life. In line with these findings, compared to normalweight subjects, obese patients display reduced GH half-life, frequency of GH secretory episodes and daily GH production rate (42). Thus, the hyposomatotropininism of obesity appears to be accounted for by at least two main abnormalities, i.e. a dramatic impairment of GH production (about 4-fold lower than that reported in normal subjects) and a significant reduction of GH half-life (about 11 minutes, compared to 15 minutes measured in normalweight volunteers). The stimulated somatotropin secretion is also markedly blunted in obesity. Indeed, impaired GH responses to both suprapituitary pharmacological stimuli (insulin-induced hypoglycaemia, arginine, galanin, L-dopa, clonidine, corticosteroids) and GHRH are documented (43). Interestingly, pretreatment with GHRH is unable to increase the somatotropin response to its subsequent application (44). On the contrary, the indirect cholinergic agonist pyridostigmine, although unable by itself to evoke a significant GH rise in obese patients (45), significantly improves the GH response to GHRH (46), and the same holds true for other compounds likely acting via inhibition of hypothalamic SRIH release (arginine, galanin, atenolol). However, though improved, the GH response is not normalized, being superimposable to that evoked in normal subjects by GHRH alone and well below the one observed in these latter after the combined stimulus. GHS represent the most powerful stimulators of GH release in obesity: both hexarelin and GHRP-6 elicit in these patients GH responses greater than those evoked by GHRH, but still lower than those registered in lean subjects (47). Accordingly, even the response to the strongest GH-releasing stimulus known in obesity, i.e. the combined administration of GHRH and GHRP-6, is significantly lower in overweight patients than in normal subjects (48). As for the peripheral limb of the somatotropin-somatomedin axis, measurements of the free fraction of IGF-I have recently allowed to demonstrate higher serum levels of this parameter in obese patients compared to lean subjects. This is likely due to the well documented hyperinsulinemia-dependent reduction of IGFBP-1 and IGFBP-2 (26). The greater availability of free, biologically active, IGF-I might explain the normal growth of obese children in spite of markedly blunted GH secretion. Obese patients display high circulating levels of GHBP (26), whose concentrations are positively correlated with BMI, and decreased binding of IGF-I to its receptor. Thus, based on the high GHBP serum levels and the reduced number of IGF-I receptors, obesity can be regarded as a condition of increased sensitivity to GH and resistance to IGF-I; this pattern is opposite to the one described for anorexia nervosa. Among the metabolic factors affecting GH secretion, the influence of blood glucose changes has been extensively studied in obesity: on one side, the GH response to hypoglycaemia is impaired; on the other side, hyperglycaemia fails to inhibit spontaneous and stimulated GH secretion (49), suggesting an inability of glucose to trigger hypothalamic SRIH release. FFA are known to negatively feedback on GH secretion. Several studies have evaluated the effects of lowering the circulating FFA levels, generally high in obesity, on GH secretion. Both acute and prolonged administration of acipimox, a nicotinic acid analogue able to block lipolysis and thus to lower serum FFA concentrations, have been shown to significantly improve spontaneous and stimulated GH release in obese subjects (50); interestingly, acipimox is able to increase the GH responsiveness even to GHRH+GHRP-6, the strongest GH stimulus known so far in obesity (51). As regards caloric restriction, a 4-day fasting does not improve the GH responsiveness to GHRH or GHRH+arginine in obese patients, whereas a severely hypocaloric diet applied for a relatively short period and unaccompanied by important weight loss significantly improves GHRH-induced but not spontaneous GH release. Of note, a more important weight reduction is followed by restoration of normal stimulated and spontaneous GH secretion in obese patients. Several interpretations have been proposed to explain the GH hyposecretion of obesity. The hypothesis of an increased central SRIH tone, suggested by the improvement of the GH response to GHRH induced by concomitant administration of drugs thought to inhibit SRIH release (pyridostigmine, arginine, galanin, atenolol), is challenged by the lack of normalization of this response even after complete suppression of SRIH by these drugs. Furthermore, we had the opportunity to demonstrate low cerebrospinal fluid levels of SRIH in obese patients (52). This observation, suggesting decreased SRIH release in the central nervous system, might envisage an adaptive response to the diminished secretion of GH and/or GHRH. A defective GHRH secretion has also been hypothesized in obese subjects based on the low frequency of spontaneous GH secretory episodes, but this possibility is questioned by the normal plasma and liquoral GHRH concentrations found in the same patients (52). The lack of GH responsiveness to repetitive GHRH administration is also in contrast with the hypothesis of GHRH deficiency, rather suggesting a functional failure of the somatotropes. In any case, this abnormality does not appear to be irreversible, given the restoration of normal GH secretion after adequate weight loss. More, the clear-cut GH rise elicited by GHS in obesity might envisage a defective secretion of the putative endogenuos ligand for GHS receptors. The elevated FFA levels might also play a role in the GH hyposecretion of obesity, as suggested by the marked improvement of somatotropin release induced by acipimox administration. Finally, the hypothesis of an enhanced negative feedback by high IGF-I levels is supported by the recent demonstration of a significant increase in free IGF-I in obesity. The marked impairment of GH secretion in obesity and the inherent protein anabolic and lipolytic properties of GH have formed the rationale for investigating, in obese patients, the effect of rhGH administration as an adjuvant of hypocaloric diets, known to cause loss of lean as well as of fat tissue. In a number of studies and in our own experience (53) the administration of rhGH at adequate doses displayed beneficial effects, i.e. sparing and enhancement of energy metabolism of lean body mass (LBM), even in severely calorie-restricted obese patients. Whether compounds less expensive than GH and able to promote its endogenous secretion may be beneficial in the therapy of weight reduction is currently under investigation.

CONCLUSIONS

The profound impact of nutritional status on GH secretion takes place through central (changes in the hypothalamic neuroendrocrine pathways regulating somatotropin release) and peripheral (macronutrients, IGF-I, insulin) influences. The alterations of the GH-IGF-I axis induced by food deprivation are common to most mammals (decreased IGF-I production, enhanced GH release) with the exception of rats (depression of the somatotropin-somatomedin axis). On the other side, overweight is invariably associated with a marked impairment of spontaneous and stimulated GH release. In any case, the abnormalities associated with both malnourishment and overnutrition are reversible after restoration of normal weight, thus likely representing an adaptive phenomenon. In particular, in chronic undernutrition the high GH with low IGF-I and insulin levels might favour lipolysis making FFA available to peripheral tissues (16), whereas in obesity the blunted GH secretion might limit the enlargement of the pool of adipocyte precursor cells available for lipid deposition and prevent a further increase in the already elevated plasma FFA which are known to worsen glucose metabolism (43). As for leptin, a consistent body of evidence suggests that the *ob* gene product promotes GH release in rodents. In these animals fasting leads to a decrease in both leptin and GH secretion while leptin-deficient or leptin-resistant obesity is associated with impaired GH release. This is in sharp contrast with the pattern displayed by humans in whom malnourishment is associated with low leptin serum concentrations and increased GH secretion, whereas overweight is characterized by high leptin levels and impaired somatotropin release. These discrepancies envisage a completely different regulation of GH secretion by leptin in humans and rodents. Alternatively, the association of GH hyposecretion and hyperleptinemia in human obesity could fit in well with the concept of a leptin resistance in this condition. The availability of leptin for in vivo human studies will help to clarify this issue.

REFERENCES

1. Giustina A, Veldhuis JD. Pathophysiology of the neuroregulation of growth hormone secretion in experimental animals and the human. Endocr Rev 1998;19:717-797.
2. Kennedy JA, Nicolson R, Wellby ML. The effect of oleic acid on the secretion of thyrotrophin and growth hormone by cultured rat anterior pituitary cells. J Endocrinol 1994;143:557-564.
3. Carroll PV, Christ ER, Bengtsson BA, et al. Growth hormone deficiency in adulthood and the effects of growth hormone replacement: a review. Growth Hormone Research Society Scientific Committee. J Clin Endocrinol Metab 1998;83:382-395.
4. Bengtsson BA, Brummer RJ, Eden S, Bosaeus I. Body composition in acromegaly. Clin Endocrinol 1989;30:121-130.
5. Considine RV, Sinha MK, Heiman ML, et al. Serum immunoreactive-leptin concentrations in normal-weight and obese women. N Engl J Med 1996;334:292-295.
6. Leal-Cerro A, Considine RV, Peino R, et al. Serum immunoreactive-leptin levels are increased in patients with Cushing's syndrome. Horm Metab Res 1996;28:711-713.
7. Florkowski CM, Collier GR, Zimmet PZ, Livesey JH, Espiner EA, Donald RA. Low-dose growth hormone replacement lowers plasma leptin and fat stores without affecting body mass index in adults with growth hormone deficiency. Clin Endocrinol 1996;45:769-773.
8. Casanueva FF, Dieguez C, Popovic V, Peino R, Considine RV, Caro JF. Serum immunoreactive leptin concentrations in patients with anorexia nervosa before and after partial weight recovery. Biochem Mol Med 1997;60:116-120.

9. Gotoda T, Manning BS, Goldstone AP, et al. Leptin receptor gene variation and obesity: lack of association in a white British male population. Hum Mol Genet 1997;6:869-876.
10. Montague CT, Farooqi IS, Whitehead JP, et al. Congenital leptin deficiency is associated with severe early-onset obesity in humans. Nature 1997;387:903-908.
11. Zhang Y, Proenca R, Maffei M, Barone M, Leopold L, Friedman JM. Positional cloning of the mouse obese gene and its human homologue. Nature 1994;372:425-432.
12. Lee GH, Proenca R, Montez JM, et al. Abnormal splicing of the leptin receptor in diabetic mice. Nature 1996;379:632-635.
13. Carro E, Senaris R, Considine RV, Casanueva FF, Dieguez C. Regulation of in vivo growth hormone secretion by leptin. Endocrinology 1997;138:2203-2206.
14. Hardie LF, Guilhot N, Trayhurn P. Regulation of leptin production in cultured mature white adipocytes. Horm Metab Res 1996;28:685-689.
15. Okada K, Sugihara H, Minami S, Wakabayashi I. Effect of parenteral administration of selected nutrients and central injection of gamma-globulin from antiserum to neuropeptide Y on growth hormone secretory pattern in food-deprived rats. Neuroendocrinology 1993;57:678-686.
16. Thissen JP, Ketelslegers JM, Underwood LE. Nutritional regulation of the insulin-like growth factors. Endocr Rev 1994;15:80-101.
17. Sugihara H, Emoto N, Shibasaki T, Minami S, Wakabayashi I. Increased pituitary growth hormone-releasing factor (GRF) receptor messenger ribonucleic acid expression in food-deprived rats. Brain Res 1996;742:355-358.
18. Tannenbaum GS, Lapointe M, Gurd W, Finkelstein JA. Mechanisms of impaired growth hormone secretion in genetically obese Zucker rats: roles of growth hormone-releasing factor and somatostatin. Endocrinology 1990;127:3087-3095.
19. Ahmad I, Finkelstein JA, Downs TR, Frohman LA. Obesity-associated decrease in growth hormone releasing-hormone gene expression: a mechanism for reduced growth hormone mRNA levels in genetically obese Zucker rats. Neuroendocrinology 1993;58:332-337.
20. Clemmons DR, Klibanski A, Underwood LE, et al. Reduction of plasma immunoreactive somatomedin-C during fasting in humans. J Clin Endocrinol Metab 1981;53:1247-1250.
21. Merimee TJ, Zapf J, Froesch ER. Insulin-like growth factors in the fed and fasted states. J Clin Endocrinol Metab 1982;55:999-1002.
22. Ho KY, Veldhuis JD, Johnson ML, et al. Fasting enhances growth hormone secretion and amplifies the complex rhythms of growth hormone secretion in man. J Clin Invest 1988;81:968-975.
23. Hartman ML, Veldhuis JD, Johnson ML, et al. Augmented growth hormone (GH) secretory burst frequency and amplitude mediate enhanced GH secretion during a two-day fast in normal men. J Clin Endocrinol Metab 1992;74:757-765.
24. Hartman ML, Clayton PE, Johnson ML, et al. A low dose euglycemic infusion of recombinant human insulin-like growth factor I rapidly suppresses fasting-enhanced pulsatile growth hormone secretion in humans. J Clin Invest 1993;91:2453-2462.
25. Ho PJ, DeMott Friberg R, Barkan AL. Regulation of pulsatile growth hormone secretion by fasting in normal subjects and patients with acromegaly. J Clin Endocrinol Metab 1992;75:812-819.
26. Argente J, Caballo N, Barrios V, et al. Multiple endocrine abnormalities of the growth hormone and insulin-like growth factor axis in prepubertal children with exogenous

obesity: effect of short- and long-term weight reduction. J Clin Endocrinol Metab 1997;82:2076-2083.

27. Counts DR, Gwirtsman H, Carlsson LM, Lesem M, Cutler GB. The effect of anorexia nervosa and refeeding on growth hormone-binding protein, the insulin-like growth factors (IGFs), and the IGF-binding proteins. J Clin Endocrinol Metab 1992;75:762-767.
28. Brambilla F, Ferrari E, Cavagnini F, et al. α2-adrenoceptor sensitivity in anorexia nervosa: GH response to clonidine or GHRH stimulation. Biol Psychiatry 1989;25:256-264.
29. Brambilla F, Lampertico M, Sali L, et al. Clonidine stimulation in anorexia nervosa: growth hormone, cortisol, and beta-endorphin responses. Psychiatry Res 1987;20:19-31.
30. Sizonenko PC, Rabinovitch A, Schneider P, Paunier L, Wollheim CB, Zahnd G. Plasma growth hormone, insulin, and glucagon responses to arginine infusion in children and adolescents with idiopathic short stature, isolated growth hormone deficiency, panhypopituitarism, and anorexia nervosa. Pediatr Res 1975;9:733-738.
31. Nakagawa K, Matsubara M, Obara T, Kubo M, Akikawa K. Responses of pituitary and adrenal medulla to insulin-induced hypoglycemia in patients with anorexia nervosa. Endocrinol Jpn 1985;32:719-724.
32. Sherman BM, Halmi KA. Effect of nutritional rehabilitation on hypothalamic-pituitary function in anorexia nervosa. In: Vigersky RA, ed. Anorexia nervosa. New York: Raven Press; 1977:211-223.
33. Scacchi M, Invitti C, Pincelli AI, Pandolfi C, Dubini A, Cavagnini F. Lack of growth hormone response to acute administration of dexamethasone in anorexia nervosa. Eur J Endocrinol 1995;132:152-158.
34. Scacchi M, Pincelli AI, Caumo A, et al. Spontaneous nocturnal growth hormone secretion in anorexia nervosa. J Clin Endocrinol Metab 1997;82:3225-3229.
35. Scacchi M, Veldhuis JD, Pincelli AI, Tomasi P, Delitala G, Cavagnini F. Evidence for a disordered mode of growth hormone secretion in anorexia nervosa. Proc 80th Meeting of The Endocrine Society, New Orleans, LA, 1998:162.
36. Stoving RK, Veldhuis JD, Flyvbjerg A, et al. Jointly amplified basal and pulsatile growth hormone (GH) secretion and increased process irregularity in women with anorexia nervosa: indirect evidence for disruption of feedback regulation within the GH-insulin-like growth factor I axis. J Clin Endocrinol Metab 1999;84:2056-2063.
37. Gerner RH, Yamada T. Altered neuropeptide concentrations in cerebrospinal fluid of psychiatric patients. Brain Res 1982;238:298-302.
38. Giusti M, Foppiani L, Ponzani P, Cuttica CM, Falivene MR, Valenti S. Hexarelin is a stronger GH-releasing peptide than GHRH in normal cycling women but not in anorexia nervosa. J Endocrinol Invest 1997;20:257-263.
39. Pincelli AI, Scacchi M, Danesi L, Dubini A, Bonfanti S, Cavagnini F. ACTH, cortisol- and GH-releasing effect of hexarelin in anorexia nervosa. Proc 80th Meeting of The Endocrine Society, New Orleans, LA, 1998:159.
40. Pincelli AI, Scacchi M, Caumo A, et al. The growth hormone - insulin-like growth factor I axis in anorexia nervosa: spontaneous GH secretion and GH-induced IGF-I generation. International symposium on growth hormone Basic aspects and new clinical applications. Marbella: Lilly Press, 1998.
41. Masuda A, Shibasaki T, Hotta M, Suematsu H, Shizume K. Study on the mechanism of abnormal growth hormone (GH) secretion in anorexia nervosa: no evidence of involvment of a low somatomedin-C level in the abnormal GH secretion. J Endocrinol Invest 1988;11:297-302.

42. Veldhuis JD, Iranmanesh A, Ho KKY, Waters MJ, Johnson ML, Lizarralde G. Dual defects in pulsatile growth hormone secretion and clearance subserve the hyposomatotropism of obesity in man. J Clin Endocrinol Metab 1991;72:51-59.
43. Scacchi M, Pincelli AI, Cavagnini F. Growth hormone in obesity. Int J Obes 1999;23:260-271.
44. Ghigo E, Procopio M, Maccario M, et al. Repetitive GHRH administration fails to increase the response to GHRH in obese subjects . Horm Metab Res 1993;25:305-308.
45. Tagliaferri M, Scacchi M, Dubini A, Cavagnini F. Effect of acute administration of pyridostigmine and pirenzepine on ACTH, growth hormone and cortisol secretion in Cushing's and in obese patients. Neuroendocrinol Lett 1994;16:145-150.
46. Cordido F, Casanueva FF, Dieguez C. Cholinergic receptor activation by pyridostigmine restores growth hormone (GH) responsiveness to GH-releasing administration in obese subjects: evidence for hypothalamic somatostatinergic participation in the blunded GH release of obesity. J Clin Endocrinol Metab 1989;68:290-293.
47. Cordido F, Penalva A, Dieguez C, Casanueva FF. Massive growth hormone (GH) discharge in obese subjects after the combined administration of GH-releasing hormone and GHRP-6: evidence for a marked somatotroph secretory capability in obesity. J Clin Endocrinol Metab 1993;76:819-823.
48. Cordido F, Penalva A, Peino R, Casanueva FF, Dieguez C. Effect of combined administration of growth hormone (GH)-releasing hormone, GH-releasing peptide-6, and pyridostigmine in normal and obese subjects. Metabolism 1995;44:745-748.
49. Maccario M, Procopio M, Grottoli S, et al. In obesity the somatotrope response to either growth hormone-releasing hormone or arginine is inhibited by somatostatin or pirenzepine but not by glucose. J Clin Endocrinol Metab 1995;80:3774-3778.
50. Andreotti AC, Lanzi R, Manzoni MF, Caumo A, Moreschi A, Pontiroli AE. Acute pharmacologic blockade of lipolysis normalizes nocturnal growth hormone levels and pulsatility in obese subjects. Metabolism 1994;43:1207-1213.
51. Cordido F, Peino R, Penalva A, Alvarez CV, Casanueva FF, Dieguez C. Impaired growth hormone secretion in obese subjects is partially reversed by acipimox-mediated plasma free fatty acid depression. J Clin Endocrinol Metab 1996;81:914-918.
52. Brunani A, Invitti C, Dubini A, et al. Cerebrospinal fluid and plasma concentrations of SRIH, beta-endorphin, CRH, NPY and GHRH in obese and normalweight subjects. Int J Obes 1995;19:17-21.
53. Tagliaferri M, Scacchi M, Pincelli AI, et al. Metabolic effects of biosynthetic growth hormone treatment in severely energy-restricted obese women. Int J Obes 1998;22:836-841.

16
EFFECTS OF GROWTH HORMONE ON SODIUM HOMEOSTASIS

Grace Y. W. Kam, Troels Wolthers, Ken K. Y. Ho
Pituitary Research Unit, Garvan Institute of Medical Research,
St. Vincent's Hospital, Sydney, Australia

INTRODUCTION

Total body water constitutes about 60% of total body weight and of this, 55% is intracellular and 45% extracellular. The extracellular fluid compartment is further divided into plasma, interstitial fluid, lymph, and transcellular fluids. Since sodium (Na) is the predominant extracellular cation, regulation of extracellular volume (ECV) is tightly linked to adjustments of body Na content. The kidneys are the major organ responsible for fluid and Na homeostasis. Glomerular filtration rate (GFR) and renal plasma flow (RPF) have direct effects on the filtered load of Na, while subsequent tubular Na reabsorption is regulated chiefly by the renin-angiotensin-aldosterone system. Other hormones including atrial natriuretic peptide and antidiuretic hormone also play important roles in fluid and Na homeostasis. Growth hormone (GH) is the most abundant pituitary hormone in the adult pituitary gland. After cessation of longitudinal growth following puberty, GH continues to play a vital role in the regulation of substrate metabolism and body composition in adult life. Direct metabolic effects of GH include the stimulation of lipolysis, fat oxidation and hepatic glucose output. The anabolic actions of GH are mediated by insulin-like growth factor I (IGF-I) which promotes protein synthesis and reduces protein oxidation. Another well-known effect of GH is fluid and Na retention. Bierring and Nielsen in 1922 first reported fluid retention in rats given anterior pituitary extracts (1). The same phenomenon was observed during administration of human GH (hGH) in man by Ikkos *et al (2)* and later confirmed by other investigators (3-7). Further research was restricted by limited supply of hGH and the concern of possible contamination of hGH extracts by unknown bioactive substances. The availability of recombinant hGH, together with the emergence of new assays and metabolic investigative tools in the past decade, have advanced the understanding of the fluid and Na regulatory effects of GH. This chapter will review the physiological mechanisms behind the antinatriuretic and antidiuretic effects of GH, as well as the associated therapeutic implications on GH administration.

GH PHYSIOLOGY & PATHOPHYSIOLOGY

The physiological and pathophysiological effects of GH on Na homeostasis have been studied in several different models, most commonly examining patients with acromegaly or GH deficiency (GHD), and investigating the effects of short-term GH administration. Most studies evaluating the effects of GH on body fluid compartments employ dilutional techniques using radioactive Na or bromide to estimate ECV, and radiolabelled albumin to estimate plasma volume (PV).

ECV based on measurement of total body Na or exchangable Na (8-10), and PV (11, 12) are increased in acromegaly. The magnitude of ECV and PV are significantly related to the logarithm of GH level. Treatment of acromegaly by adenectomy (9) or long-acting somatostatin analogue octreotide results in reduction in PV. It has been well established for many years that fat free mass, which comprises ECV, body cell mass, and bone, is increased in acromegaly. Combining Na dilution and dual energy X-ray absorptiometry, O'Sullivan *et al* (13) have demonstrated that the increase in fat free soft tissue mass, i.e. fat free mass not inclusive of bone, in acromegaly can be accounted for almost entirely by expansion of ECV (Figure 1). Moreover, the increase in ECV was reversible with octreotide, identifying GH as the agent by which these changes were induced.

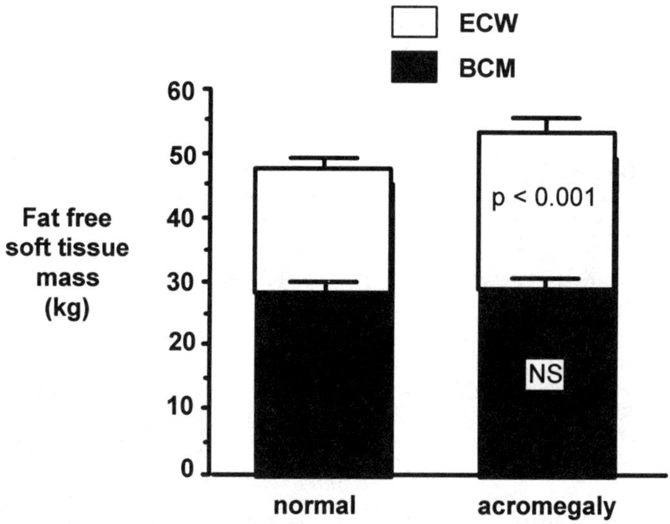

Figure 1. Comparison of fat free soft tissue mass in acromegalics and normal subjects, showing constituents of extracellular water compartment (ECW) and body cell mass (BCM). *P* value was determined by Student's paired *t*-test. Data are shown as mean ± SEM.

In contrast, GHD patients are characterised by diminished ECV and PV (14, 15). Hoffman *et al* (16) demonstrated reduction in fat free soft tissue mass in GHD patients compared to normal subjects, which corresponded to proportionate decreases in ECV and body cell mass (Figure 2). Ho & Weissberger (17) observed significant reduction in both 24 hour urinary Na excretion and urine volume in normal subjects given 5 days of rhGH at a dose of 0.2 IU/kg/day. Serum Na concentration transiently fell on day 3 of GH administration, while urine osmolarity was not altered throughout the study. Similarly, Moller et *al* (18) studied the effects of 6-day of rhGH administered at a dose of 6 $IU/m^2/day$ in normal subjects and reported a transient reduction in 24 hour urinary Na on day 2, which was no longer significant by day 6 of treatment. Replacement of GH in GHD patients resulted in significant ECV expansion as early as by day 3 and normalisation of ECV by the 3rd week, whereas PV increased only after 3 weeks of treatment (19).

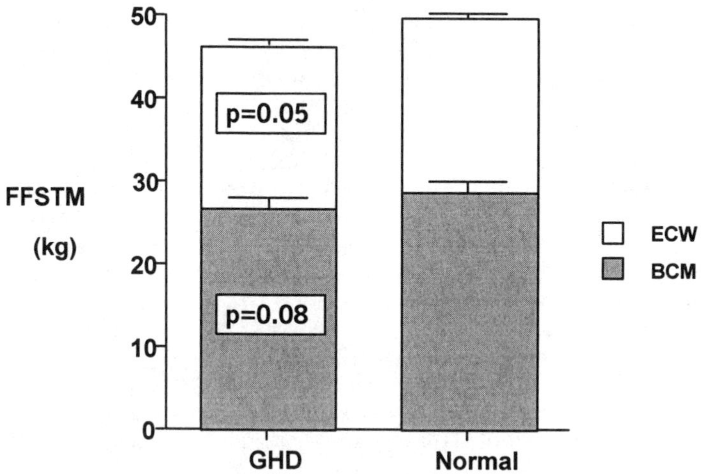

Figure 2. Comparison of fat free soft tissue mass (FFSTM) in GHD and normal subjects, showing constituents of extracellular water compartment (ECW) and body cell mass (BCM). *P* value was determined by Student's paired *t*-test. Data are shown as mean ± SEM.

Thus, GH treatment increases ECV acutely in GHD patients, whereas a longer period is required to normalise both ECV and PV. In summary, GH causes Na and fluid retention leading to expansion of the ECV and PV.

MECHANISMS OF SODIUM & FLUID RETENTION

Renin-angiotensin-aldosterone system

Renin is secreted by the juxtaglomerular apparatus in the kidneys in response to fluid or Na depletion. It cleaves angiotensinogen from the liver to angiotensin I, which is then converted to angiotensin II by angiotensin converting enzyme in vascular endothelium especially in the lungs. Angiotensin II not only increases reabsorption of Na at the proximal renal tubules (20), but also stimulates the adrenal cortex to produce aldosterone, resulting in enhanced distal tubular reabsorption of water and Na.

The renin-angiotensin system was the first to be considered and studied as a possible mediator of the antinatriuretic effect of GH. Early studies reported increased or unchanged urinary aldosterone excretion in man after GH administration (5-7), while animal studies in rats and dogs treated with hGH revealed increased aldosterone synthesis (21, 22). More recent studies of GH administration to GHD patients and normal subjects demonstrated enhanced plasma renin activity (17, 19, 23) with peak levels at day 3-5 (17). Plasma aldosterone was also found to be increased (17, 24) or unchanged (19, 23) upon GH administration.

Hoffman *et al* (19) observed an increase in angiotensinogen level and a trend towards increasing angiotensin II and aldosterone levels in GHD patients replaced with rhGH after 7 days. Wyse *et al* (25) observed upregulation of angiotensin II receptors in the liver, kidneys and adrenals of dwarf rats treated with GH. A dipsogenic effect of angiotensin II could account for the increased thirst sensation reported during the first 2-3 days of GH administration (17). Moreover, increase in angiotensinogen level in GHD patients after GH replacement (19) could be due to GH stimulation of hepatic production of angiotensinogen. Thus, activation of the renin-angiotensin-aldosterone system is likely to underlie in part the antinatriuretic effect of GH.

Recently, Moller *et al* (18) conducted a randomised, double-blind, placebo-controlled study to evaluate the role of renin-angiotensin system in GH-induced fluid retention. Eight young normal male subjects received in random order four 6-day treatment periods with placebo, GH alone, or coadministered with either captopril or spironolactone. Evidence for activation of renin-angiotensin system by GH was provided by the finding of an increase in plasma renin activity and angiotensin II levels during all 3 GH treated periods as compared to placebo. GH alone induced ECV expansion and reduction in bioimpedance, the latter being an indirect biophysical marker of fluid retention. These effects were completely abolished by blockade of the renin-angiotensin system at different levels by captopril and spironolactone (Figure 3). These data provide strong evidence that the effects of GH on fluid homeostasis are largely mediated through activation of the renin-angiotensin-aldosterone system.

However, with chronic GH replacement in GHD children, no increase in plasma renin activity (26) or plasma aldosterone level (26, 27) could be detected after 6 and 12 months. This observation suggests that activation of the renin-angiotensin system by GH is an early event occurring during the initial weeks of GH administration and failure to detect significant changes during chronic GH replacement therapy may result from homeostatic adaptation of the renin-angiotensin system to increased body fluid and Na load (28).

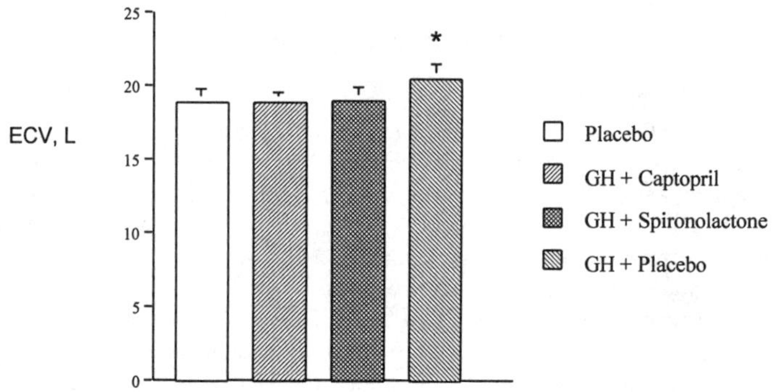

Figure 3. Comparison of extracellular volume (ECV) after 6-days treatment with placebo, GH and captopril, GH and spironolactone, and GH and placebo. $*P < 0.05$ *vs* placebo.

Direct Renal Effects of GH

GH may also act directly on the renal tubules to induce Na retention. Evidence for this notion comes from the observation of Na retention during hGH administration in bilaterally adrenalectomized animals (27) and human subjects (7). GH has significant effects on renal haemodynamics. GFR and renal plasma flow (RPF) are elevated in acromegaly (11) and reduced after pituitary surgery (29,30). GFR is also reduced in hypopituitarism (31) and is increased after GH administration in both normal subjects (32) and GHD patients (33). These changes in GFR and RPF may be direct effects of GH or indirect via IGF-I actions (see below). In general, an increase in GFR is accompanied by proportional increase in tubular fluid and Na reabsorption resulting in no net change in Na excretion. However, the observation that GH induces a net reduction in urinary Na excretion suggests that the increase in tubular Na reabsorption is over and above that resulting from concomitant increase in GFR. This direct tubular effect of GH is supported by the identification of GH receptors in basolateral plasma membrane of isolated proximal tubules (34).

Renal Effects of IGF-I

Elevation in GFR and RPF following GH may be mediated via effects of IGF-I on the kidneys. Hirshberg *et al* (35) reported that after a single intramuscular injection of GH, elevation in GFR and RPF occurred after 24 hours and persisted until day 3. This paralleled the time course of serum IGF-I increments. Guler *et al* (36) infused rIGF-I intravenously in normal subjects for 5 days and observed a 25-30% increase in GFR and RPF. But in contrast to the 2 kg weight gain seen in GH-treated subjects (37), no weight gain resulted from IGF-I infusion. These results suggest that GH and IGF-I may have distinct but complementary actions on the kidney, with IGF-I increasing GFR and RPF and GH causing fluid and Na retention in the renal tubules.

Cardiac Effects of GH

GH exerts both anabolic and inotropic effects on the heart. GH administration in GHD patients increases cardiac mass (38-40), reverses diastolic dysfunction (40), and elevates cardiac index (41). The improved cardiac output can be translated clinically into enhanced RPF and thus increased GFR. On the other hand, expansion of ECV and PV by GH improves venous return to the heart, ultimately increasing cardiac output. The effects of GH on the cardiovascular system may therefore influence and be influenced by fluid and Na homeostasis. A detailed description of the cardiac effects of GH is beyond the scope of this review and is dealt with in greater detail by other contributors elsewhere in this monograph.

Atrial Natriuretic Peptide

Atrial natriuretic peptide secretion is stimulated by atrial distension (42) and volume expansion (43). It brings about natriuresis by increasing GFR and reducing Na reabsorption (44). There is evidence that GH suppresses secretion of atrial natriuetic peptide. Acromegalic patients have reduced responsiveness of atrial natriuretic peptide to saline infusion (45). Moreover, in normal subjects (15, 17, 24), the administration of GH reduced atrial natriuretic peptide levels despite an increase in ECV. These findings suggest that the reduced response in atrial natriuretic peptide secretion to ECV expansion following GH administration may contribute to the fluid and Na retaining properties of GH.

Insulin

Fluid and Na retention seen following GH administration may also be mediated in part by insulin. Studies undertaken over 20 years ago demonstrated that insulin infusion resulted in acute retention of Na, occurring due to increased reabsprtion at the distal renal tubules (46-48). Since GH stimulates insulin secretion and causes sustained hyperinsulinaemia, the role of insulin as a possible mediator of GH-induced Na retention cannot be excluded. Furthermore, Na retention observed during IGF-I infusion may be exerted via stimulation of insulin receptors by elevated levels of free IGF-I. This is unlike the situation during GH administration, where complexed IGF-I is increased due to concomitant increase in IGF binding protein 3 (IGFBP-3) and acid labile subunit (ALS).

Antidiuretic Hormone

Antidiuretic hoemone is released from the posterior pituitary gland in response to stimulation of osmoreceptors in the hypothalamus by increased serum Na level or osmolarity. It increases the permeability of distal renal tubules and collecting ducts, thereby enhancing water reabsorption. Antidiuretic hormone secretion was observed to be unchanged rather than suppressed, as would have been expected with volume expansion, in normal subjects given GH (17, 24). This finding suggests that antidiuretic hormone, in line with atrial natriuretic peptide and insulin, may be indirectly involved in GH related Na retention.

THERAPEUTIC IMPLICATIONS

Fluid retention is a significant component of soft tissue swelling in acromegaly, and is reversible with treatment that effectively reduces GH. In clinical practice, symptoms such as hand swelling and carpel tunnel syndrome rapidly resolve in the immediate postoperative period following adenectomy or during the early phase of octreotide treatment. Recent understanding of the pathological mechanism underlying GH induced Na and fluid retention suggests that spironolactone may be an effective treatment to alleviate symptoms of fluid retention in refractory cases of acromegaly. Strangely, such therapeutic usage involving blockade of the renin-angiotensin system has not been evaluated systematically to date.

The benefits of GH replacement in GHD adults have been unequivocally demonstrated (49). However, side-effects relating to fluid retention such as arthralgia and myalgia, carpel tunnel syndrome, peripheral oedema, weight gain or even hypertension were reported in up to 50% of the subjects who received GH in the early trials (33, 50-52). The dosage of GH used was adopted from the experience with GHD children, and was well in excess of the normal daily adult GH production rate of approximately 0.02-0.03 IU/kg/day (53). Supranormal IGF-I levels were frequently observed with excessive GH doses. A recent multicentre trial (54) employing a fixed dosage regimen reported peripheral oedema in 40% and supranormal IGF-I levels in 35% of GHD adults treated with GH. As IGF-I is a reliable marker of GH action, more recent studies employing replacement regimens which titrated GH dosages to maintain IGF-I in the normal range, reported virtually no occurrence of excessive fluid retention (55, 56). The doses of GH used in these latter studies were about 30-40% lower than those used in earlier studies.

Recently, Hoffman *et al* (19) examined the dose-response relationship between GH and markers of fluid retention in GHD patients. Normalisation of IGF-I was achieved using GH dosage of 0.04 IU/kg/day, whereas higher GH dosage of 0.08 IU/kg/day resulted in supranormal IGF-I. Moreover, significant increase in body weight and ECV only occurred during supraphysiological GH treatment (Figure 4). Moller *et al* (57) and De Boer *et al* (58) also achieved normalisation of IGF-I, IGFBP-3 and ALS using similar range of GH dosage.

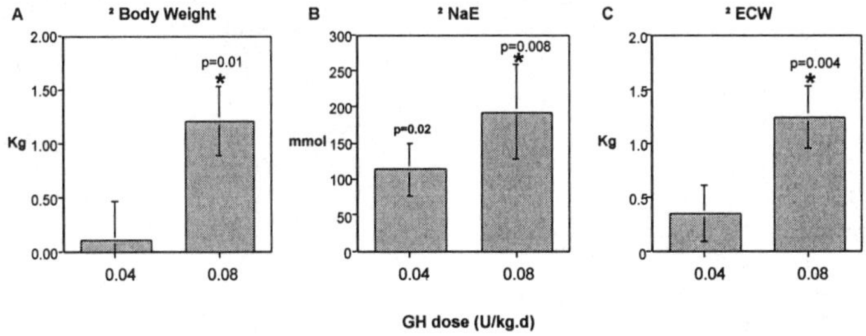

Figure 4. Mean (± SEM) changes from placebo in body weight (BW) (A), exchangable solium (NaE) (B), and extracellular water (ECW) (C) during low (0.04 μg/kg/day) and high doses (0.08 μg/kg/day) of GH treatment.

Fluid and Na retention is dose-dependent and an expected action of GH. It is therefore incorrect to describe it as a side-effect of physiological GH replacement therapy. Symptomatic fluid retention generally occurs when excessive GH doses are used. The most appropriate regimen therefore includes a low starting dose followed by careful and slow escalation of dose while maintaining IGF-I within age-matched normal range.

CONCLUSION

GH regulates fluid and Na homeostasis, which is one of the many physiological properties of GH. Fluid retention is not a side-effect of GH, but an expected consequence of biological action of GH. Replacement of GH in GHD patients restores contracted ECV and PV to normal. This is achieved indirectly via activation of the renin-angiotensin system and alteration in other hormones which influence Na homeostasis, and by direct renal effects of GH and possibly IGF-I. In clinical practice, symptoms of fluid retention may serve as a useful clinical guide for monitoring GH replacement dosage in GHD adults.

ACKNOWLEDGMENTS

Dr. Grace Y. W. Kam was a recipient of an overseas medical training fellowship from the Hospital Authority of Hong Kong. This work was supported in part by a grant from the National Health and Medical Research Council of Australia.

REFERENCES

1. Bierring E, Nielsen E. The composition of the tissues of albino rats treated with alkaline anterior pituitary extracts. Biochem J 1932;26:1015-1021.
2. Ikkos D, Luft R, Gemzell CA. The effect of human growth hormone in man. Acta Endocrinol (Copenh) 1959;32:341-361.

3. Hutchings JJ, Escamilla RF, Deamer WC, Li CH. Metabolic changes produced by human growth hormone in a pituitary dwarf. J Clin Endocrinol Metab 1959;19:759-769.
4. Henneman PH, Forbes AP, Moldawer M, Dempsey EF, Carroll EL. Effects of human growth hormone in man. J Clin Invest 1960;37:1223-1245.
5. Beck JC, McGarry EE, Dyrenfurth I, Venning EH. Metabolic effects of human and monkey growth hormone in man. Science 1957;125:884-885.
6. Venning EH, Dyrenfurth VL, Giroud JP, Beck JC. Effect of growth hormone on aldosterone secretion. Metabolism 1956;5:697-702.
7. Biglieri EG, Watlington CO, Forsham PH. Sodium retention with human growth hormone and its subfractions. J Clin Endocrinol Metab 1961;21:361-370.
8. Aloia JF, Roginsky MS, Jowsey J, Dombrowski CS, Shukla KK, Cohn SH. Skeletal metabolism and body composition in acromegaly. J Clin Endocrinol Metab 1972;35:543-551.
9. McLellan AR, Connell JMC, Beastall GH, Teasdale G, Davies DL. Growth hormone, body composition and somatomedin C after treatment in acromegaly. Q J Med 1988;69:997-1008.
10. Ritchie CM, Sheridan B, Fraser R, et al. Studies on the pathogenesis of hypertension in Cushing's disease and acromegaly. Q J Med 1990;76:855-867.
11. Falkheden T, Sjogren B. Extracellular fluid volume and renal function in pituitary insufficiency and acromegaly. Acta Endocrinol (Copenh) 1964;46:80-88.
12. Strauch G, Lego A, Therain F, Bricaire H. Reversible plasma and red blood cell volume increases in acromegaly. Acta Endocrinol (Copenh) 1977;85:465-478.
13. O'Sullivan AJ, Kelly JJ, Hoffman DM, Freund J, Ho KKY. Body composition and energy expenditure in acromegaly. J Clin Endocrinol Metab 1994;78:381-386.
14. Rosen T, Bosaeus I, Tolli J, Lindstedt G, Bengtsson B-A. Increased body fat mass and decreased extracellular fluid volume in adults with growth hormone defiency. Clin Endocrinol 1993;38:63-71.
15. Moller J, Frandsen E, Fisker S, Jorgensen JOL, Christiansen JS. Decreased plasma and extracellular volume in growth hormone deficient adults and the acute and prolonged effects of GH administration: a controlled experimental study. Clin Endocrinol 1996;44:533-539.
16. Hoffman DM, O'Sullivan AJ, Freund J, Ho KKY. Adults with growth hormone deficiency have abnormal body composition but normal energy metabolism. J Clin Endocrinol Metab 1995;80:72-77.
17. Ho KKY, Weissberger AJ. The antinatriuretic action of biosynthetic human growth hormone in man involves activation of the renin-angiotensin system. Metabolism 1990;39:133-137.
18. Moller J, Moller N, Frandsen E, Wolthers T, Jorgensen JOL, Christiansen JS. Blockade of the renin-angiotensin-aldosterone system prevents growth hormone-induced fluid retention in humans. Am J Physiol 1997;272:E 803-808.
19. Hoffman DM, Crampton L, Sernia C, Nyugen TV, Ho KKY. Stort term growth hormone (GH) treatment of GH-deficient adults increases body sodium and extracellular water, but not blood pressure. J Clin Endocrinol Metab 1996;81:1123-1128.
20. Harris PJ, Navar LG. Tubular transport responses to angiotensin. Am J Physiol 1985;248:F 621-630.
21. Venning EH, Lucius OJ. Effect of growth hormone on the biosynthesis of aldosterone in the rat. Endocrinology 1962;20:486-491.

22. Cushman P, Alter S, Hilton JG. Effects of various growth hormone preparations on dog adrenal cortical secrtetion. Endocrinology 1966;78:971-975.
23. Cuneo RC, Saloman F, Wilmshurst P, et al. Cardiovascular effects of growth hormone treatment in growth-hormone-deficient adults: stimulation of the renin-aldosterone system. Clin Sci 1991;81:587-592.
24. Moller J, Jorgensen JOL, Moller N, Hansen KW, Pedersen EB, Christiansen JS. Expansion of extracellular volume and suppression of atrial natriuretic peptide after growth hormone administration in normal man. J Clin Endocrinol Metab 1991;72:768-772.
25. Wyse B, Waters M, Sernia C. Stimulation of the renin-angiotensin system by growth hormone in Lewis dwarf rats. Am J Physiol 1993;265:E 332-339.
26. McCaa RE, Montalvo JM, McCaa CS. Role of growth hormone in the regulation of aldosterone biosynthesis. J Clin Endocrinol Metab 1978;46:247-253.
27. Ludens JH, Bach RR, Williamson HE. Characteristics of the antinatriuretic action of growth hormone. Proc Soc Exp Biol Med 1969;130:1156-1158.
28. Kelly JJ.1993 in Garvan Institute of Medical Research, pp. 16, University of New South Wales, Sydney, Australia
29. Falkheden T. Renal function following hypophysectomy in man. Acta Endocrinol (Copenh) 1963;42:571-590.
30. Falkheden T, Wickbom I. Renal function and size following hypophysectomy in man. Acta Endocrinol (Copenh) 1965;48:348-354.
31. Luft R, Sjogren B. The significance of the adenohypophysis, adrenal cortex, and thyroid in renal function in man. Acta Endocrinol (Copenh) 1950;4:351-362.
32. Christiansen JS, Gammelgaard J, Orskov H, Andersen AR, Telmer S, Parving H-H. Kidney function and size in normal subjects before and during growth hormone administration for one week. Eur J Clin Invest 1981;11:487-490.
33. Jorgensen JO, Pedersen SA, Thuesen L, et al. Beneficial effects of growth hormone treatment in GH-deficient adults. Lancet 1989;1:1221-1225.
34. Rogers SA, Hammerman MR. Growth hormone activates phopholipase C in proximal tubular basolateral membranes from the canine kidney. Proc Natl Acad Sci USA 1989;86:6363-6366.
35. Hirschberg R, Kopple JD. Effects of growth hormone on GFR and renal plasma flow in man. Kidney Int 1987;32 (Suppl 22):S 21-24.
36. Guler H, Eckardt K, Zapf J, Bauer C, Froesch ER. Insulin-like growth factor-1 increases glomerular filtration rate and renal plasma flow in man. Acta Endocrinol (Copenh) 1989;121:101-106.
37. Guler H, Schmid C, Zapf J, Froesch ER. Effect of recombinant insulin-like growth factor-1 on insulin secretion and renal function in normal human subjects. Proc Natl Acad Sci USA 1989;86:2868-2872.
38. Cuneo RC, Salomon FS, Wilmshurst P, et al. Cardiovascular effects of growth hormone treatment in growth-hormone-deficient adults: stimulation of the renin-aldosterone system. Clin Sci 1991;81:587-592.
39. Amato G, Carella C, Fazio S, et al. Body composition, bone metabolism, and heart structure and function in growth hormone (GH)-deficient adults before and after GH replacement therapy at low doses. J Clin Endocrinol Metab 1996;77:1671-1676.
40. Valcavi R, Gaddi O, Zini M, Iavicoli M, Mellino U, Portioli I. February Cardiac performance and mass in adults with hypopituitarism: effects of one year of growth hormone treatment. J Clin Endocrinol Metab 1995;80:659-666.

41. Thuesen L, Jorgensen JOL, Muller JR, et al. Short and long-term cardiovascular effects of growth hormone therapy in growth hormone deficient adults. Clin Endocrinol 1994;41:615-620.
42. Lang RE, Tholken H, Ganten D, Luft FC, Ruskoaho H, Unger T. Atrial natriuretic factor-a circulating hormone stimulated by volume loading. Nature 1985;314:264-266.
43. Weil J, Lang RE, Suttmann H, Rampf U, Bidlingmaier F, Gerzer R. Concomitant increase in plasma atrial natriuretic peptide and cyclic GMP during volume loading. Klin Wochenschr 1985;63:1265-1268.
44. Dunn BR, Ichikawa I, Pfeffer JM, Troy JL, Brenner BM. Renal and systemic haemodynamic effects of synthetic atrial natriuretic peptide in the anaesthetised rat. Circ Res 1986;59:237-246.
45. McKnight JA, McCance DR, Hadden DR, et al. Basal and saline-stimulated levels of plasma atrial natriuretic factor in acromegaly. Clin Endocrinol 1989;31:431-438.
46. De Fronzo RA, Cooke CR, Andres R, Faloona GR, Davis PJ. The effect of insulin on renal handling of sodium, potassium, calcium, and phosphate in man. J Clin Invest 1975;55:845-855.
47. De Fronzo RA, Goldberg M, Agus ZS. The effects of glucose and insulin on renal electrolyte transport. J Clin Invest 1976;58:83-90.
48. De Fronzo RA. The effect of insulin on renal sodium metabolism. Diabetologia 1981;21:165-171.
49. Carroll PV, Christ ER, Members of Growth Hormone Research Society Scientific Committee, et al. Growth hormone deficiency in adulthood and the effects of growth hormone replacement: a review. J Clin Endocrinol Metab 1998;83:382-395.
50. Salomon F, Cuneo RC, Hesp RC, Sonksen PH. The effects of treatment with recombinant human growth hormone on body composition and metabolism in adults with growth hormone deficiency. N Engl J Med 1989;321:1797-803.
51. Whitehead HM, Boreham C, McIlrath EM. Growth hormone treatment of adults with growth hormone deficiency: results of a 13-month placebo controlled cross-over study. Clin Endocrinol 1992;36:45-52.
52. Binnerts A, Swart GR, Wilson JHP, et al. The effect of growth hormone administration in growth hormone deficient adults on bone, protein, carbohydrate and lipid homeostasis, as well as on body composition. Clin Endocrinol 1992;37:79-87.
53. Jorgensen JOL. Human growth hormone replacement therapy: pharmacological and clinical aspects. Endocr Rev 1991;12:189-208.
54. Cuneo RC, Judd S, Wallace JD, et al. The Australian multicenter trial of GH treatment in GH-deficient adults. J Clin Endocrinol Metab 1998;83:107-116.
55. Drake WM, Coyte D, Camacho-Hubner C, et al. Optimizing growth hormone replacement therapy by dose titration in hypopituitary adults. J Clin Endocrinol Metab 1998;83:3913-3919.
56. Baum HBA, Biller BMK, Finkelstein JS, et al. Effects of physiologic growth hormone therapy on bone density and body composition in patients with adult-onset growth hormone deficiency. Annals of Internal Medicine 1996;125:883-890.
57. Moller J, Jorgensen JOL, Lauersen T, et al. Growth hormone dose regimens in adult GH deficiency: effects on biochemical growth markers and metabolic parameters. Clin Endocrinol 1993;39:403-8.
58. De Boer H, Blok GJ, Popp-Snijders C, Stuurman L, Baxter RC, van der Veen E. Monitoring of growth hormone replacement therapy in adults, based on measurement of serum markers. J Clin Endocrinol Metab 1996;81:1371-1377.

17
GROWTH HORMONE AND EXERCISE PERFORMANCE

William B. Wehrenberg, Lisa K. Conley,
Filippo Manelli^, Andrea Giustina^
College of Agriculture, Forestry and Life Sciences, Clemson University, Clemson – South Carolina;
^Endocrine Section, Dept of Internal Medicine, University of Brescia, Italy

INTRODUCTION

Growth hormone (GH) is a potent anabolic and lipolityc agent that also affects carbohydrate metabolism (1). The anabolic actions of GH are evident in GH deficient children who start to grow rapidly when GH is administered (2). In adults the anabolic nature of GH is dramatically exhibited by patients with acromegaly, and indirectly by subjects with GH deficiency (GHD) (3,4). The exercise is an important factor, as well as age, gender and nutrition, that regulates GH secretion (5). Therefore, it should be of no surprise that GH, and its tissue mediator IGF-I, in turn appear to have important roles in the physiologic response to exercise. No where is this latter notion better supported than in exercise-related studies conducted in individuals who exhibit altered GH tone. We now turn our attention to the effects of elevated and decreased circulating GH concentrations on exercise performance and to the effects of GH replacement or supplementation on exercise performance in GHD and in normal subjects.

ABNORMAL CIRCULATING GH LEVELS AND EXERCISE PERFORMANCE

Elevated Circulating GH Concentrations. Chronically elevated circulating concentrations of GH and IGF-I are the hallmark of acromegaly (6). This prolonged increase in blood GH and IGF-I concentrations has profound effects on the anatomy and the function of many systems. Most notably, acromegalic individuals exhibit: proliferation of cartilage and other connective tissues, and thus, enlarged hands and feet, altered joint dysfunction and distorted facial features; hyperglycemia with insulin insensitivity, often leading to diabetes mellitus and its inherent complications (3), and profound changes in cardiac structure and function that can eventually lead to congestive heart failure (7). With regard to the last point, these cardiac specific changes include interstitial fibrosis and lymphomononuclear infiltration, ventricular wall thickening (hypertrophy) with decreased end-diastolic wall stress, abnormal ventricular diastolic filling and reduced ventricular ejection fractions (8).
The duration of elevated circulating GH concentrations appears to be a major determinant with regard to impairment of cardiac performance as it appears that shorter disease durations result in lesser degrees of impairment, especially in younger individuals (6). Acromegalics exhibit decreased exercise tolerance. For example, it has been reported (9) that patients with

active acromegaly are only able to perform about half as much work at both their anaerobic threshold and maximal exercise abilities than normal controls matched for age and sex. Impaired cardiac function is most likely one of the primary contributing factors. Researchers have reported significantly reduced left and right ventricular ejection fractions in exercising acromegalics as compared to normal exercising controls (10). It should be noted, however, that O'Keefe (11), observed no impairment of left ventricular function either at rest or during exercise in their patient population unless cardiac disease (e.g., coronary artery disease) other than hypertrophy was also present. These discordant results may be reconciled if one consider that the atherosclerotic cardiac disease itself in the second study may have resulted from active acromegaly, and the possibility that the subjects in the different studies may have suffered from the disease for different durations. Thus, a great number of systems most likely are involved in the decreased exercise tolerance noted in acromegalic individuals. It appears that decreasing circulating GH and IGF-I concentrations in this patient population may improve cardiac performance and, therefore, exercise tolerance. Octreotide, the first somatostatin analog clinically available, has been shown in long- and short-term studies to decrease cardiac hypertrophy and improve diastolic function in acromegaly (12). It has been demonstrated that a 24-h continuous infusion of octreotide is able to improve workload and oxygen consumption during a cycloergometric exercise in patients with acromegaly and cardiac hypertrophy (13) both at anaerobic threshold and at maximal exercise. Moreover, 6 months administration of s.c. octreotide improves exercise performance in acromegalic patients (14). Recently, it also has been demonstrated a significant decrease in mean GH values during slow release Lanreotide (SR-L) treatment (15). Moreover, the administration of SR-L causes beneficial and persistent effects on cardiac function. In fact, single SR-L injection caused an improvement in the cardiopulmonary response to exercise at the anaerobic threshold after both 7 and 14 days from the injection (16). Particularly, the endocrine (decrease in GH levels) and echocardiographic responses to SR-L are maximal after 7 days from the injection. Overall, the effect of somatostatin analogs on the exercise performance in acromegaly could be hypothesized to be due to a reduction of the inotropic action of GH which, exerted on hypertrophic myocardium, is likely to result in negative hemodynamic effect.

Decreased Circulating GH Concentrations. GH deficiency (GHD) in adults is characterized by increased visceral adiposity, lipidemia and increased occurrence of atherosclerotic disease, as well as decreased muscle mass and bone mineral density, lowered body water content and depressed cardiac function (17). With regard to cardiac function and in contrast to what is seen in acromegalic subjects, patients with GHD exhibit decreased left ventricular wall thickness with a concomitant increased end-systolic wall stress, and decreased cardiac output (18). Collectively, changes in many systems manifest themselves as complaints of general weakness and fatigue and, most strikingly, decreased exercise tolerance. As with acromegaly, the decreased exercise tolerance seen in GHD is characterized by low maximum workloads achieved and tolerance of exercise durations of only short duration (19). The impaired cardiac function noted at rest is exaggerated with exercise, with researchers reporting decreased maximal heart rates, depressed left ventricular systolic and diastolic function, and reduced ejection fractions with respect to exercising controls (20). Changes in other systems also appear to play a role in the decreased exercise tolerance of GHD patients. In summary, adults with GHD have signs of premature aging, in terms of decreased muscle and cardiovascular function, increased body fat and decreased bone mineral density. Treatment with GH significantly improves these abnormalities. Before GH replacement, exercise capacity, determined by stepwise increasing the work load on a bicycle ergometer until exhaustion (power output), was reduced in adults with GHD. A normal consequence of aging is a decrease in the overall function of the growth axis, and therefore, in circulating GH and IGF-I concentrations (5). It has been suggested that the age-related changes in such parameters as body composition (increased percentage of fat; decreased lean muscle mass (21) and bone strength (lessened due to decreased bone density) (22) may be a consequence of this gradual decline in GH concentrations (23). The importance of GH in aging processes remains controversial (24). For example, aging is accompanied by a decreases in muscle strength. The act of exercising itself and not GH supplementation appears to have a greater impact on improving muscle strength in older adults. By analogy with the fall in ovarian function that inevitably eventually occurs in women with increasing age, this fall in GH secretion has been termed the somatopause. It is established that a number of the clinical features of the somatopause are shared with the syndrome of adult-onset GHD and it has been proposed the importance of GH in maintaining health and vitality with increasing age (23). Considering the profound effects that severe GHD (e.g., resulting from classically defined hypopituitarism) has on cardiovascular function, it would appear that study of the impact of age-related declines in circulating GH concentrations on cardiac performance both at rest and during exercise is of importance. Clearly, more studies investigating the importance of GH in the exercising elderly are warranted.

GH ADMINISTRATION AND EXERCISE PERFORMANCE

GHD patients. Numerous studies have shown that GH replacement significantly increased exercise performance, determined as power output or as exercise time, which normalized completely after 38 months of GH treatment (25). Maximal VO2, as well as anaerobic threshold increased significantly after GH treatment, whereas maximal heart rate was found to be either unchanged or increased following GH replacement (4). Particularly, GH replacement therapy improves cardiac function, muscle mass, bone strength and, therefore, exercise tolerance in GHD subjects (26,27). Six months of GH replacement therapy in hypopituitary adults produced significant improvements their in exercise capacity (28). This may be related to improved cardiac output and systolic performance resulting from increases in ventricular wall thickness and stress. Longer-term treatment appears to bring more beneficial effects. In fact, it has been shown that 3-5 years of GH replacement therapy in a cohort of men with childhood onset GHD resulted in a 16% increase in stroke volumes and a 33% increase in cardiac outputs, without a maintained increase in left ventricular masses (29). In addition, increases in lower extremity muscle mass and bone mineral density were noted following the GH therapy. Collectively, these improvements resulted in a substantial increase (over pretreatment values) in the maximal work loads these subjects were able to sustain (29). In conclusion, GH therapy in GHD improves exercise performance, which has mainly been attributed to the increased muscle mass, but also an increased cardiac output during exercise has been proposed as important for the beneficial effect of GH on physical capacity.

Normal subjects. Few studies have documented the effects of acute and more prolonged GH administration in normal adult subjects (30,31).
In the young men, resistance exercise with or without GH therapy for 12 week resulted in similar increments in muscle size, strength, and muscle protein synthesis, indicating that: 1) the increase in fat-free mass with GH treatment was probably due to an increase in lean tissue other than skeletal muscle; 2) resistance training supplemented with GH did not further enhance muscle anabolism and function. In 1989, GH replacement was shown to be beneficial in adults with GH deficiency, and in 1990 Rudman showed remarkably similar beneficial effects in a group of elderly men with low plasma IGF-I values, but no underlying pituitary pathology, who were administered GH (23). In these adults, low doses of GH increased lean body mass and bone mineral density, decreased body fat and lowered LDL cholesterol. More recently, to investigate whether age-related deficits in the somatotropic axis limit the degree to which muscle strength can improve with resistance training in older individuals, a supplementation with rhGH were taken in elderly men without a significant augment in the response to strength training (32). These results suggested that deficits in GH secretion do not underlie the time-dependent leveling off of muscle strength seen with training in the elderly and provide no support for the popular view of GH as an ergogenic aid. Other observations suggested that resistance exercise training improved muscle strength and anabolism in older men, but these improvements were not enhanced when exercise was combined with daily GH administration (33). The greater increase in fat-free mass with GH treatment may have been due to an increase in noncontractile protein and fluid retention. Research regarding the impact of GH treatment on other aspects of exercise is at present sparse (34).

Athletes. Although there is no scientific evidence documenting an improvement in athletic performance following GH supplementation, it is reported that this practice is becoming widespread among athletes wishing to avoid detection with current doping control measures (35). There are anedoctal reports that athletes are injecting cadaveric or biosynthetic forms of GH, both of which are associated with potentially serious complications. In addition, some athletes are ingesting drugs and amino acids in the belief that their endogenous GH secretion will be increased. There have been no scientific studies on the effects of GH supplementation, and the anedoctal reports have been equivocal, with some individuals reporting spectacular results while others report no change. In a study using GH treatment in power athletes compared with a placebo-control group, the results indicated no increase in maximal strength during concentric contraction of the biceps and quadriceps muscles, although levels of IGF-I were doubled (36). In highly trained power athletes with low fat mass and high lean body mass, no additional effect of GH treatment on strength is to be expected. In conclusion, it is clear that the anthropometric ramifications, especially with respect to muscle mass, of the metabolic actions of GH and IGF-I treatment in normal subjects and atheltes require further study. At present, it appears that daily GH or IGF-I treatment modestly increases nitrogen retention in most normal adults, probably by separate but permissive mechanisms, but only for a short period of time (approximately 1 month) (37). During prolonged GH administration, resistance to the anabolic actions of GH seems to occur, and optimizing the anabolic effects of GH or IGF-I treatment will require a better understanding of the interactions among GH, GHBP, IGF-I production, IGFBPs, the GH dose regimen, and other unidentified regulatory factors. On the basis of the similar increases in muscle protein synthesis, muscle cross-sectional area, and muscle strength observed in placebo and GH-treated exercising young adults, it is doubtful that the nitrogen retention associated with daily GH treatment results in an increase in contractile protein, improved muscle function, strength and athletic performance. Further, the side effects of GH treatment (water retention, carpal tunnel syndrome, insulin resistance) would be a detriment, rather than an aid, to athletic performance. In addition, whether prolonged (> 6 months) GH treatment alone or in combination with other agents used by athletes is associated with other adverse side effects (e.g., cancer, diabetes) has not been evaluated. Therefore, health professionals should continue to discourage the use of GH by exercise enthusiasts (38). The introduction anti-doping of methods able to detect this practice may also contribute to minimize the use of GH doping treatment (39).

CONCLUSIONS

GH is an important anabolic agent in children as well as in adults (40). The anabolic properties of GH have been investigated extensively. The effects of GH on normal, hypertrophied and atrophied muscles have been studied previously in animal experiments that demonstrated an increase in muscle weight and size, but no comparable increase in performance or tension (41). In children with GH deficiency, Turner's syndrome or intrauterine growth retardation, an increase in muscle tissue is seen after treatment with GH (42). In adults with GH deficiency, the changes in body composition can be corrected by GH treatment; lean body mass and strength increase within a few months (43). Nevertheless, the available data on GHD support the concept that GH therapy alone, in the absence of some form of exercise program, may increase the amount of lean tissue but not the quality of functional capacity of this tissue and it may be that training, in addition to GH therapy, may be necessary to significantly increase physical performance in these patients (44). In acromegalics with long-standing GH hypersecretion, the muscle volume is increased, but muscle strength and performance are not improved (45). These observations gave rise to the interest shown by healthy subjects and athletes in using GH to increase their muscle mass and strength. The improvements in muscle strength obtained by resistance exercise training in healthy older men or young men were not enhanced by additional administration of GH (31,32). The larger increases in fat-free mass observed in the GH-treated groups were obviously not due to accretion of contractile protein, but rather to fluid retention or accumulation of connective tissue (33).

The clandestine use of ergogenic aids in various aspects of sports has been of concern for quite some time (35). GH's efficacy of action to alter body composition such as increasing lean muscle mass (32) and to increase muscle size (40), taken with the availability of recombinant human GH that is hard to detect in drug screening, make illegal use of this pharmacological agent quite tempting to many individuals. This is even though the increased muscle size noted in response to exogenous GH treatment does not necessarily translate into improved strength (44). GH supplementation without GH deficiency is also quite troubling, aside from the moral issue of fairness in competition, if one considers the devastating consequences of chronically elevated circulating GH concentrations (3). As the discussion above points out, and although a direct connection has not been sufficiently established, such prolonged GH elevation runs the risk of acromegaly-like pathology with: 1) altered glucose utilization with insulin intolerance that eventually leads to diabetes mellitus (3) and 2) severe cardiac anatomical and functional pathology that can eventually lead to heart failure (8). It has also been suggested that the development of GH antibodies might be a concern, especially when one considers that this occurrence could result in functional GH deficiency (a pathological state with its own set of health consequences discussed below)

(35). It is doubtful that most individuals without GH deficiency who are considering using GH supplementation to augment their performance and to improve their appearance understand the potential dire physiological consequences of this practice.

In summary, the most consistent effect of GH observed during both short- and long-term GH administration appears to be mediated by a stimulation of protein synthesis rather than an inhibition of proteolysis as well as cardiac performance (33). However, it does appear evident that training in adults is far more important for maintaining or increasing skeletal muscle mass and cardiopulmonary efficacy, which is not probably amplified by GH administration (33).

REFERENCES

1. Raben M. Growth hormone. N Engl J Med 1962; 266:31-35.
2. Abs R, Bengtsson BA, Hernberg-Stahl E, Monson JP, Tauber JP, Wilton P, Wuster C. GH replacement in 1034 growth hormone deficient hypopituitary adults: demographic and clinical characteristics, dosing and safety. Clin Endocrinol (Oxf) 1999;50(6):703-713.
3. Melmed S. Acromegaly. N Engl J Med 1990;22:966-967.
4. Jorgensen JOL, Pedersen SA, Thuesen L, Jorgensen J, Ingemann Hansen T, Skakkebaek NE, Christiansen JS. Beneficial effects of growth hormone treatment in GH-deficient adults. Lancet 1989;1:1221-1225.
5. Giustina A, Veldhuis JD. Pathophysiology of the neuroregulation of growth hormone secretion in experimental animals and the human. Endocr Rev 1998;19:717-797.
6. Colao A, Cuocolo A, Marzullo P, Nicolai E, Ferone D, Della Morte AM, Petretta M, Salvatore M, Lombardi G. Impact of patient's age and disease duration on cardiac performance in acromegaly: A radionuclide angiography study. J Clin Endocrinol Metab 1999;84:1518-1528.
7. Lombardi G, Colao A, Ferone D, Marzullo P, Orio F, Longobardi S, Merola B. Effect of growth hormone on cardiac function. Horm Res 1997;48:S38-S42.
8. Fazio S, Cittadini A, Sabatini D, Merola B, Colao A, Biondi B, Lombardi G, Sacca L. Evidence for biventricular involvement in acromegaly: A Doppler echocardiographic study. Eur Heart J 1993;14:26-33.
9. Giustina A, Zaltieri G, Negrini F, Wehrenberg WB. The pharmacological aspects of the treatment of acromegaly. Pharmacol Res 1996;34:1-22.
10. Colao A, Cuocolo A, Marzullo P, Nicolai E, Ferone D, Florimonte L, Salvatore M, Lombardi G. Effects of 1 year treatment with octreotide on cardiac performance in patients with acromegaly. J Clin Endocrinol Metab 1999;84:17-23.
11. O'Keefe JC, Grant SJ, Wiseman JC, Stiel JN, Wilmshurst EG, Cooper RA, Edwards AC. Acromegaly and the heart: echocardiographic and nuclear imaging studies. Aust N Z J Med 1982;12:603-607.
12. Pereira JL, Rodriguez-Puras MJ, Leal-Cerro A, Martinez A, Garcia-Luna PP, Gavilan I, Pumar A, Astorga R. Acromegalic cardiopathy improves after treatment with increasing dose of octreotide. J Endocrinol Invest 1991;14:17-23.
13. Giustina A, Boni E, Romanelli G, Grassi V, Giustina G. Cardiopulmonary performance during exercise in acromegaly, and effects of acute suppression of growth hormone hypersecretion with octreotide. Am J Cardiol 1995;75:1042-1047.

14. Padayatty SJ, Perrins EJ, Belchetz PE. Octreotide treatment increases exercise capacity in patients with acromegaly. Eur J Endocrinol 1996;134:554-559.
15. Morange-Ramos PCI, Cogne M, Jaquet P. Three years follow-up of acromegalic patients treated with intramuscolar slow-release lanreotide: J Clin Endocrinol Metab 1997;82:18-22.
16. Manelli F, Desenzani P, Boni E, Bugari G, Negrini F, Romanelli G, Grassi V, Giustina A. Cardiovascular effects of a single slow release lanreotide injection in patients with acromegaly and left ventricular hypertrophy. Pituitary 1999;2:205-210.
17. Cuneo RC, Salomon F, McGauley GA, Sonksen PH. The growth hormone deficiency syndrome in adults. Clin Endocrinol (Oxf.) 1992;37:387-397.
18. Fazio S, Cittadini A, Sabatini D, Merola B, Colao A, Biondi B, Longobardi S, Lombardi G, Sacca L. Growth hormone and heart performance. A novel mechanism of cardiac wall stress regulation in humans. Eur Heart J 1997;18:181-184.
19. Cittadini A, Cuocolo A, Merola B, Fazio S, Sabatini D, Nicolai E, Colao A, Longobardi S, Lombardi G, Sacca L. Impaired cardiac performance in growth hormone deficient adults and its improvement after growth hormone replacement. Am J Physiol 1994;267:E219-E225.
20. Cuocolo A, Nicolai E, Colao A, Longobardi S, Cardei S, Fazio S, Merola B, Lombardi G, Sacca L, Salvatore M. Improved left ventricular function after growth hormone in patients with hypopituitarism. Assessment with radionuclide angiography. Eur J Nucl Med 1996;23:390-394.
21. Forbes GB, Reina JC. Adult lean body mass declines with age: Some longitudinal observations. Metabolism 1970;19:653-663.
22. Riggs BL, Wahner HW, Dunn WL, Mazess RB, Offord KP, Melton LJ. Differential changes in bone mineral density of the appendicular and axial skeleton with aging. J Clin Invest 1981;67:328-335.
23. Rudman D, Feller AG, Nagraj HS, Gergans GA, Lalitha PY, Goldberg AF, Schlenker RA, Cohn L, Rudman IW, Mattson DE. Effects of human growth hormone in men over 60 years old. N Engl J Med 1990;323:1-6.
24. Corpas E, Harman SM, Blackman MR. Human growth hormone and human aging. Endocrine Rev 1993;14:20-39.
25. Jorgensen JOL, Thuesen L, Muller J, Ovesen P, Skakkebaek NE, Christiansen JS. Three years growth hormone treatment in growth hormone deficient adults. Eur J Endocrinol 1994;130:224-228.
26. Cuneo RC, Salomon F, Wiles CM, Hesp R, Sonksen PH. Growth hormone treatment in growth hormone deficient adults. I. Effects on muscle mass and strength. J Appl Physiol 1991;70:688-694.
27. Cuneo RC, Salomon F, Wiles CM, Hesp R, Sonksen PH. Growth hormone treatment in growth hormone deficient adults. III. Effects on exercise performance. J Appl Physiol 1991;70:695-700.
28. Jorgenson JOL, Pedersen SA, Thuesen L, et al. Beneficial effects of growth hormone treatments in GH_deficient adults. Lancet 1989;1:1221-1225.
29. Ter Maaten JC, De Boer H, Kamp O, Stuurman L, Van Der Veen EA. Long-term effects of growth hormone (GH) replacement in men with childhood-onset GH deficiency. J Clin Endocrinol Metab 1999;84:2373-2380.
30. Rudman D, Kutner MH, Rogers CM, Lubin MF, Fleming A, Bain R. Impaired growth hormone secretion in the adult population. J Clin Invest 1981;67:1361-1369.

31. Yarasheski KE, Campbell JA, Smith K, Rennie MJ, Holloszy JO, Bier DM. Effect of growth hormone and resistance exercise on muscle growth in young men. Am J Physiol 1992;262:E261-267.
32. Taaffe DR, Pruitt L, Reim J, Hintz RL, Butterfield G, Hoffman AR, Marcus R. Effect of recombinant human growth hormone on the muscle strength response to resistance exercise in elderly men. J Clin Endocrinol Metab 1994;79:1361-1366.
33. Yarasheski KE, Zachwieja JJ, Campbell JA, Bier DM. Effect of growth hormone and resistance exercise on muscle growth and strength in older men. Am J Physiol 1995;268:E268-276.
34. Savine R, Sonksen PH. Is the somatopause an indication for growth hormone replacement? J Endocrinol Invest 1999;22:142-149.
35. Macintyre JG. Growth hormone and athletes. Sport Med 1987;4:129-142.
36. Frisch H. Growth hormone and body composition in athletes. J Endocrinol Invest 1999;22:106-109.
37. Yarasheski KE. Growth hormone effects on metabolism, body composition, muscle mass, and strength. Exerc Sport Sci Rev 1994;22:285-312.
38. Clarkson PM, Thompson HS. Drugs and sport. Research findings and limitations. Sports Med 1997;24(6):366-384.
39. Wu Z, Bidlingmaier M, Dall R, Strasburger CJ. Detection of doping with human growth hormone. Lancet 1999;13:895.
40. Sacca L, Cittadini A, Fazio S. Growth hormone and the heart. Endocr Rev 1994;15:555-573.
41. Bigland B, Jehring B. Muscle performance in rats, normal, and treated with growth hormone. J Physiol 1952;129:136.
42. Salomen F, Cuneo RC, Hesp R, Sonksen PH. The effects of treatment with recombinant human growth hormone on body composition and metabolism in adults with growth hormone deficiency. N Engl J Med 1989;321:1797-1803.
43. Salomon F, Cuneo RC, Hesp R, Sonksen PH. The effects of treatment with recombinant human growth hormone on body composition and metabolism in adults with growth hormone deficiency. N Engl J Med 1989;321:1797-1803.
44. Rodriguez-Arnao J, Jabbar A, Fulcher K, Besser GM, Ross RJM. Effects of growth hormone replacement on physical performance and body composition in GH deficient adults. Clin Endocrinol 1999;51:53-60.
45. Naugelsparen M, Trickey R, Davies MJ, Jenkins JS. Muscle changes in acromegaly. Brit Med J 1976;2:914-915.

18
IGF-1 AND THE HEART

Marc Y Donath and Juergen Zapf
Division of Endocrinology and Diabetes, Dept of Medicine, University Hospital, Zurich

1) INTRODUCTION

The so-called somatomedin hypothesis (1) states that growth hormone (GH) acts on skeletal tissues indirectly by inducing the production of a growth factor, somatomedin, circulating in blood and mediating the effects of GH on growth. After the isolation and amino sequencing of insulin-like growth factor I (IGF I) and its biological characterization it became evident that this polypeptide was the postulated somatomedin. IGF I consists of 70 amino acids and has a molecular weight of 7649 (2). It is mainly produced and secreted by the liver. However, many other tissues synthesize IGF I which acts locally on these tissues. The significance of endocrine versus auto-/paracrine actions of IGF I is still under debate. Two recent publications provide evidence that in mice liver-derived circulating IGF I is not required for growth (3, 4). IGF I tissue expression is under the control of GH (5), but other factors, like insulin or nutrition, also influence IGF I production. Both endogenous and administered IGF I mimic trophic effects of pituitary GH. In contrast, some effects of administered IGF I on intermediary metabolism are opposite to those of GH (6, 7). Moreover, the two hormones act by different mechanisms, which implies that endogenous IGF I induced by GH does in general not mediate the effects of GH on intermediary metabolism.
Cardiac effects of IGF were first investigated by Meuli and Froesch in 1975 in the perfused rat heart (8). IGF, like insulin, stimulated glucose transport and lactate production. The biological potency ratio between IGF and insulin in this model was higher than in adipose tissue. Based on this finding and on competitive binding studies, it was concluded that insulin and IGF act on the heart separately via the insulin and the type 1 IGF receptor, respectively (9). In 1986 Borner and Froesch showed that IGF I stimulated the contractility of the isolated perfused rat heart (Figure 1). Since the availability of pure IGF I was limited at that time the positive inotropic effects of IGF I could not be extensively tested, and the data remained unpublished.

2) CARDIAC HYPERTROPHY

Infusion of recombinant human (rh) IGF I or rhGH into hypophysectomized rats causes differential effects on organ growth (10). In particular, IGF I stimulates heart but not skeletal muscle weight, whereas GH increases the weight of both organs. Furthermore, IGF I increases the synthesis of cardiac proteins in vitro (11, 12, 13). These observations led us to investigate the effect of IGF I on cultured adult rat cardiomyocytes (14).

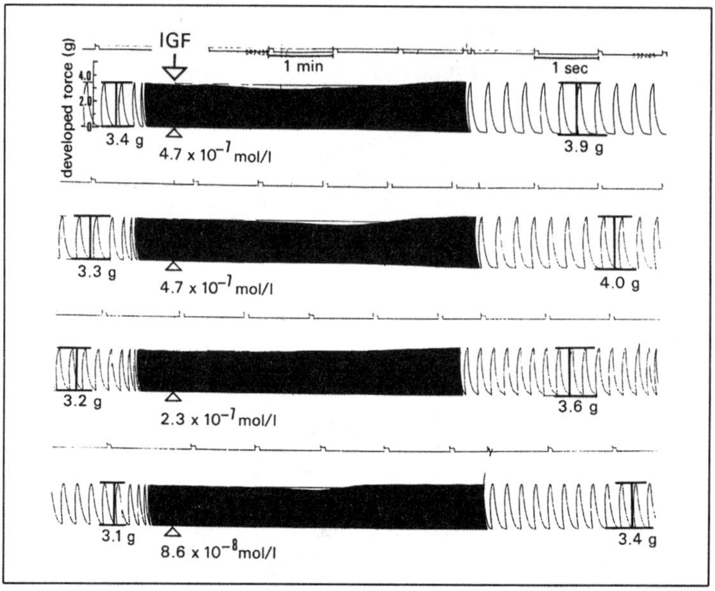

Figure 1. Effect of IGF I on contractile force of the isolated perfused rat heartIsolated rat hearts were perfused with 5 mM glucose, 2g/l of human serum albumin in Krebs-Ringer bicarbonate buffer at 37°C. Contraction amplitude was recorded on a dynograph with a tension transducer. The heart muscle was connected to a transducer unit. IGF I was tested at 3 different concentrations. (From F. Borner and E. R. Froesch, unpublished)

In long-term culture, adult rat cardiomyocytes reexpress fetal proteins, such as β- myosin heavy chain (MHC), smooth muscle-α-actin and atrial natriuretic factor (ANF). The recapitulation of this "fetal program" is also found in overload-induced cardiac hypertrophy, suggesting similarities between the *in vitro* and *in vivo* situation. IGF I-treated adult rat cardiomyocyte cultures showed a conspicuous increase in the number of newly built sarcomers (Figure 2). The increased formation of contractile elements was accompanied by a decreased expression of smooth muscle-α-actin (14) and ANF (15).
Based on these findings and on findings reported in the literature, we proposed a hypothesis to explain the development of heart hypertrophy *in vivo* resulting from pressure/volume overload, and subsequent recompensation promoted by IGF I action (Figure 3, (14)). According to this hypothesis, overload increases wall stress, the stimulus for hypertrophy, during a first phase. Subsequently, transforming growth factor-β1, basic fibroblast growth factor and catecholamines may participate in transforming the increased mechanical load on the heart into biochemical signals (16, 17, 18, 19, 20). During a second phase, these signals induce the production of scaffold proteins, such as α-actin-isoforms, for sarcomerogenesis (21, 22). At the same time, they cause a shift from the α- to the β-MHC as the main isoform expressed (23), and lead to up-regulation of ANF (24, 25, 26), thus initiating diuresis, and of IGF I (27, 28, 29, 30, 31) as well as of type 1 IGF receptors (14, 32).

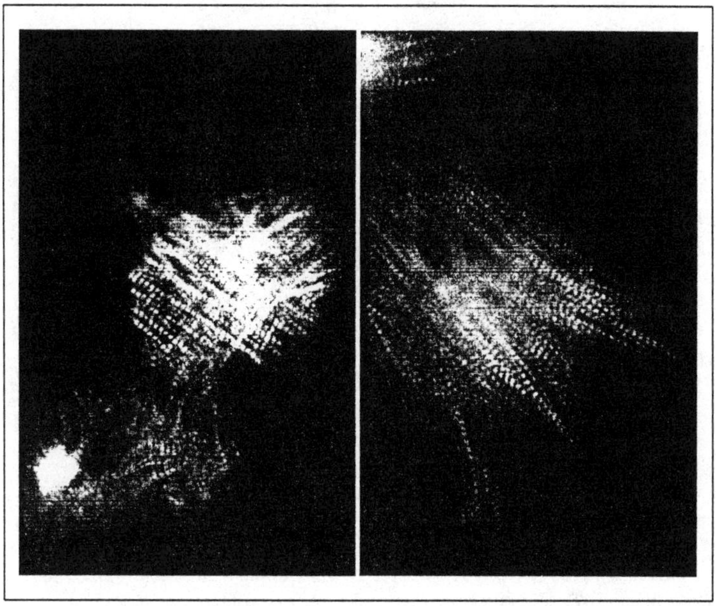

Figure 2. Immunofluorenscense labeling of cardiac C-protein of adult rat cardiomyocytes. Cardiomyocytes were maintained in culture for 16 days. Left, untreated cells. Right, typical rhIGF I-treated (500 ng/ml) cell with myofibrils across the whole cell body (x360). (From ref. 14, with permission)

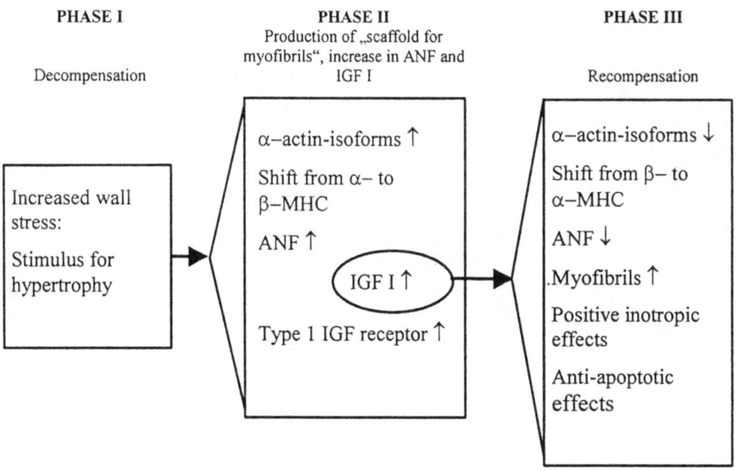

Figure 3. Hypothetical model for the development of hypertrophy resulting from pressure/volume overload, and subsequent recompensation.

During a third phase, locally produced IGF I may act in an autocrine/paracrine manner to promote the formation of new myofibrils and to down-regulate α-actin-isoforms which are no longer needed as a scaffold, as well as to downregulate ANF whose diuretic effect is no longer required. Finally, positive inotropic and anti-apoptotic effects of IGF I (described below) may also enhance the adaptation of the heart to overload.
In order to test this hypothesis, we investigated the effect of a 6-day infusion of rhIGF I or rhGH into normal and hypophysectomized rats on the ventricular expression of myofibrillar genes and ANF (33). The results of this study confirmed that IGF-I stimulates expression of myofibrillar genes and modulates ANF expression in rat heart ventricles *in vivo,* depending on the hormonal status of the animals. In rats rendered hypertensive by constriction of the left renal artery (see below), we observed that β-MHC mRNA in the left ventricle was increased during the whole postoperative period and that IGF I considerably reduced this increase (34). Suppression of β-MHC expression results in a more "physiological" ratio of α- to β- MHC, possibly reflecting a more favorable balance between the economy of contraction and contractility, and a better readaptation to pressure overload.

3) INCREASE IN CONTRACTILITY

IGF I increases the contractility of cultured neonatal and adult rat cardiomyocytes as well as of adult dog cardiomyocytes (35, 36, 37, 38). Increased myocardial contractility by IGF I has also been demonstrated in isolated rat and ferret papillary muscles (36, 37, 39, 40) and on whole hearts in rats (39). GH had no effect on the indices of cardiac function in isolated heart preparations (38, 39). The increase of myocardial contractility observed in animals has been confirmed in humans (see below).
The mechanisms underlying the positive inotropic effect of IGF I are not clear. Freestone et al (36) observed that the positive inotropic response to IGF I is associated with a rise in free cytosolic Ca^{2+} in isolated rat cardiac myocytes. A similar observation was described by Kinugawa et al in cardiomyocytes isolated from dogs (38). However, Cittadini et al. (39) using whole rat hearts and ferret papillary muscles, reported that the positive inotropic effect of IGF I was not associated with increased intracellular Ca^{2+} availability, but rather with a increase of myofilament Ca^{2+} sensitivity. Furthermore, Ren et al (37) demonstrated that the nitric oxide synthase inhibitor nomega-nitro-L-arginine methyl ester attenuated the IGF I-induced increase in contractility of rat papillary muscles and of isolated cardiomyocytes. This suggests that, in addition, nitric oxide may mediate the stimulatory effect of IGF I on contractility.

4) PREVENTION OF APOPTOSIS

There is now ample experimental evidence that IGF I can prevent or limit programmed cell death. Thus, IGF I protects hemopoietic cells (41, 42, 43), neuronal cells (44, 45), tumor cells (46, 47, 48) and vascular smooth muscle cells (49) from undergoing apoptosis.
Buerke et al demonstrated that, IGF I administered 1 hr before myocardial ischemia limited myocardial reperfusion injury in rats, possibly by inhibition of apoptosis and of leukocyte-induced cardiac necrosis (50). Overexpression of IGF I in transgenic mice reduced myocyte apoptosis and necrosis in the region adjacent to infarcted myocardium and attenuated ventricular dilatation, myocardial loading, and cardiac hypertrophy (51). In primary cultures of fetal rat cardiomyocytes, pretreatment with IGF I 20 minutes prior to induction of

apoptosis by serum withdrawal and addition of doxorubicine, reduced the number of apoptotic cells (52). The protective effect of IGF I was associated with decreased Bax induction, suppressed caspase 3 activation and reduced Bcl2-associated Bax. Leri et al. investigated the effect of stretching on apoptosis in myocytes isolated from adult rat heart (53). Myocyte stretching activates p53 and p53-dependent genes, leading to the formation of angiotensin II and to apoptosis. When added 15 minutes in advance, IGF I attenuated these effects.

Taken together, these data show that IGF I can inhibit apoptosis in cardiomyocytes. However, in all the experiments described above, IGF I had always been added prior to the induction of apoptosis. This is different from the clinical situation where ischemia occurs before therapeutic intervention. Therefore, it remains to be investigated whether IGF I has also a protective effect if administered after induction of apoptosis.

5) IMPROVEMENT OF CARDIAC FUNCTION IN ANIMALS

Several experimental studies have shown that IGF I improves myocardial function, probably by mechanisms including stimulation of myofibrillogenesis, increase of contractility and prevention of apoptosis as described above. Treatment of normal adult rats with IGF I induces cardiac hypertrophy without fibrosis along with an increase in cardiac index and a decrease in systemic vascular resistance (54, 55, 56). In the severely dysfunctional heart following myocardial infarction similar beneficial effects have been described (55,57). Furthermore, IGF I had a cardioprotective effect in doxorubicine-treated rats (58).

We studied the effects of IGF I after induction of hypertension by constriction of the left renal artery in rats (34). In the vehicle-treated hypertensive animals body weight gain was reduced, whereas IGF I-treated hypertensive animals continued to gain weight like sham-operated animals. Left ventricular weight and the left, but not the right ventricle/body weight ratio increased more in IGF I than in vehicle-infused rats. Beta-MHC mRNA in the left ventricle was increased in hypertensive controls during the whole postoperative period, and IGF I nearly abolished this increase. On the first postoperative day, IGF I prevented the decrease of alpha- MHC mRNA and the increase of ANF mRNA in the left ventricle. The restoration of body weight gain and stimulation of left ventricular cardiac weight by IGF I as well as the partial reversion of hypertension-induced changes in cardiac protein expression may reflect enhanced readaptation of the heart to overload.

6) ACUTE CARDIOVASCULAR EFFECTS IN HUMANS

Only limited data are available on acute cardiovascular effects of IGF I in humans (Table 1). These studies show that IGF I has vasodilator effects (59, 60, 61). Russell-Jones et al have shown that an intravenous infusion of IGF I increases cardiac output and stroke volume in normal human volunteers as assessed by impedance cardiography (62).

We investigated the systemic cardiovascular effects of IGF I at rest and during exercise in 8 healthy adults (63). Individuals were randomized to receive a subcutaneous injection of 60 μg/kg of rhIGF I or saline, in a crossover, double-blind trial, with an interval of 6 days between the two injections. Cardiac function and performance were evaluated by echo-cardiography and exercise tests. IGF I improved cardiac performance, with an increase in stroke volume, in cardiac output and of the ejection fraction. Left ventricular diastolic function was not affected. Exercise tests did not show any pathologic changes in the electrocardiogram. No significant changes of heart rate or blood pressure were found during rest or exercise, nor were maximal exercise duration and peak oxygen consumption influenced by IGF I. Recently, Bisi et al have confirmed the positive inotropic effects of IGF I in healthy adults by radionuclide angiocardiography (64).

We also studied the acute hemodynamic effects of IGF I in patients with chronic heart failure (65). Eight patients were randomized to receive recombinant human IGF I, 60 μg/kg, or placebo, intravenously over 4 h in a crossover, double blind study on two consecutive days. Using heart catheterization, IGF I was shown to increase cardiac index and stroke volume index, and to decrease systemic vascular resistance, right atrial pressure and pulmonary artery wedged pressure. Mean systemic and pulmonary artery pressure, as well as heart rate and pulmonary vascular resistance were not significantly influenced. Urinary levels of noradrenaline decreased significantly during IGF I infusion, which may indicate a beneficial effect of IGF I on neurohumoral activation in patients with chronic heart failure. IGF I was well tolerated in all patients, and no pathologic electrocardiogram changes were recorded. The increase in cardiac output following IGF I treatment is probably caused by a combination of direct positive inotropic effects and peripheral vasodilatation.

Table 1. Acute cardiovascular effects of IGF I

heart rate	→
systemic vascular resistance	↓
pulmonary vascular resistance*	→
right atrial pressure*	↓
systemic artery pressure	→
pulmonary artery pressure	→
pulmonary artery wedge pressure*	↓
stroke volume	↑
cardiac output	↑
ejection fraction	↑

* Measured only in patients with chronic heart failure

7) PERSPECTIVES

In vitro and *in vivo* data demonstrate that IGF I elicits significant effects on cadiomyocytes and on the cardiovascular system. It regulates myofibrillar genes, stimulates myofibril formation and down-regulates fetal cardiac genes reexpressed during overload hypertrophy, it exerts positive inotropic effects and protects cardiomyocytes from apoptosis, and it causes vasodilation. All of these effects may contribute to improve heart perfomance. Thus, IGF I has an attractive therapeutic potential for treatment of cardiac conditions such as heart failure, cardiomyopathy and ischemic heart disease.

Several of the beneficial effects of IGF I are shared with GH which induces IGF I both locally and in the circulation. We (14) and others (13, 38, 39) have not observed direct GH effects either on cultured cardiomyocytes or on isolated heart preparations during short-term perfusions. Therefore, it appears that GH effects on the heart *in vivo* are, at least partly, mediated by IGF I produced by interstitial cells adjacent to cadiomyocytes. However, whereas GH administration (7) or oversecretion (66, 67, 68, 69) in humans leads to hyperinsulinemia, insulin resistance and hyperlipidemia, IGF I administration lowers normal and elevated serum insulin, increases insulin sensitivity (6, 7, 70, 71, 72, 73), and reduces fasting and postprandial triglyceride levels (74). Thus, in addition to its potential effects on heart performance, it reduces risk factors of cardiovascular disease. It remains to be investigated whether the observed acute cardiovascular effects of IGF I prove to be beneficial also during prolonged IGF I treatment of heart disease in humans.

REFERENCES

1. Daughaday WH, Hall K, Raben MS, et al. Somatomedin: proposed designation for sulphation factor. Nature 1972; 235(5333):107.
2. Humbel RE. Insulin-like growth factors I and II. Eur J Biochem 1990; 190:445-462.
3. Yakar S, Liu JL, Stannard B, Butler A, Accili D, Sauer B, LeRoith D. Normal growth and development in the absence of hepatic insulin-like growth factor I. Proc Natl Acad Sci USA 1999; 96(13):7324-7329.
4. Sjogren K, Liu JL, Blad K, Skrtic S, Vidal O, Wallenius V, LeRoith D, Tornell J, Isaksson OG, Jansson JO, Ohlsson C. Liver-derived insulin-like growth factor I (IGF-I) is the principal source of IGF-I in blood but is not required for postnatal body growth in mice. Proc Natl Acad Sci USA 1999; 96(12):7088-7092.
5. Gosteli-Peter MA, Winterhalter KH, Schmid C, Froesch ER, Zapf J. Expression and regulation of insulin-like growth factor-I (IGF-I) and IGF-binding protein messenger ribonucleic acid levels in tissues of hypophysectomized rats infused with IGF-I and growth hormone. Endocrinol 1994; 135:2558-2567.
6. Hussain MA, Schmitz O, Mengel A, Keller A, Christiansen JS, Zapf J, Froesch ER. Insulin-like growth factor I stimulates lipid oxidation, reduces protein oxidation, and enhances insulin sensitivity in humans. J Clin Invest 1993; 92:2249-2256.
7. Hussain MA, Schmitz O, Mengel A, Glatz Y, Christiansen JS, Zapf J, Froesch ER. Comparison of the effects of growth hormone and insulin-like growth factor I on substrate oxidation and on insulin sensitivity in growth hormone-deficient humans. J Clin Invest 1994; 94: 1126-1133.
8. Meuli C, Froesch ER. Effects of insulin and of NSILA-S on the perfused rat heart: glucose uptake, lactate production and efflux of 3-0-methyl glucose. Europ J Clin Invest 1975; 5:93-99.
9. Meuli C, Froesch ER. Binding of insulin and nonsuppressible insulin-like activity to isolated perfused rat heart muscle. Arch Biochem Biophys 1976; 177:31-38.
10. Guler HP, Zapf J, Scheiwiller E, Froesch ER. Recombinant human insulin-like growth factor I stimulates growth and has distinct effects on organ size in hypophysectomized rats. Proc Natl Acad Sci USA 1988; 85: 4889-4893.
11. Florini JR, Ewton DZ. Induction of gene expression in muscle by the IGFs, Growth Regul 1992; 2:23-29.
12. Fuller SJ; Mynett JR; Sugden PH. Stimulation of cardiac protein synthesis by insulin-like growth factors. Biochem J 1992; 282:85-90.

13. Ito H, Hiroe M, Hirata Y, Tsujino M, Adachi S, Shichiri M, Koike A, Nogami A, Marumo F. Insulin-like growth factor-I induces hypertrophy with enhanced expression of muscle specific genes in cultured rat cardiomyocytes. Circulation 1993; 87:1715-1721.
14. Donath MY, Zapf J, Eppenberger-Eberhardt M, Froesch ER, Eppenberger HM. Insulin-like growth factor I stimulates myofibril development and decreases smooth muscle alpha actin of adult cardiomyocytes. Proc Natl Acad Sci USA 1994; 91:1686-1690.
15. Eppenberger-Eberhardt M, Aigner S, Donath MY, Kurer V, Walther P, Zuppinger C, Schaub MC, Eppenberger HM. IGF-I and bFGF differentially influence atrial natriuretic factor and alpha-smooth muscle actin expression in cultured atrial compared to ventricular adult rat cardiomyocytes. J Mol Cell Cardiol 1997; 29(8):2027-2039.
16. Parker TG, Packer SE, Schneider MD. Peptide growth factors can provoke "fetal" contractile protein gene expression in rat cardiac myocytes. J Clin Invest 1990; 85:507-514.
17. Schneider MD, Parker TG. Cardiac myocytes as targets for the action of peptide growth factors. Circulation. 1990; 81:1443-1456.
18. Morgan HE, Gordon EE, Kira Y, Chua BHL, Russo LA, Peterson CJ, McDermott PJ, Watson PA. Biochemical mechanisms of cardiac hypertrophy. Annu Rev Physiol 1987; 49:533-543.
19. Parker TG, Chow K-L, Schwartz RJ, Schneider MD. Differential regulation of skeletal alpha-actin transcription in cardiac muscle by two fibroblast growth factors. Proc. Natl. Acad. Sci. USA 1990; 87:7066-7070.
20. Chien KR, Knowlton KU, Zhu H, Chien S. Regulation of cardiac gene expression during myocardial growth and hypertrophy: molecular studies of an adaptive physiologic response. FASEB J 1991; 5(15):3037-3046.
21. Black FM, Packer SE, Parker TG, Michael LH, Roberts R, Schwartz RJ, Schneider MD. The vascular smooth muscle α-actin is reactivated during cardiac hypertrophy provoked by load. J Clin Invest 1991; 88:1581-1588.
22. Schwartz K, de la Bastie D, Bouveret P, Oliviéro P, Alonso S, Buckingham M. α-skeletal muscle actin mRNAs accumulate in hypertrophied adult rat hearts. Circ Res 1986; 59:551-555.
23. Izumo S, Lompre AM, Matsuoka R, Koren G, Schwartz K, Nadal-Ginard B, Mahdavi V. Myosin heavy chain messenger RNA and protein isoform transitions during cardiac hypertrophy. Interaction between hemodynamic and thyroid hormone-induced signals. J Clin Invest. 1987; 79:970-977.
24. Lattion A-l, Michel J-B, Arnauld E, Corvol P, Sourbier F. Myocardial recruitment during ANF mRNA increase with volume overload in the rat. Am J Physiol 1986; 251:H890-H896.
25. Izumo S, Nadal-Ginard B, Mahdavi V. Protooncogene induction and reprogramming of cardiac expression produced by pressure overload. Proc Natl Acad Sci USA 1988; 85:339-343.
26. Mercadier J-J, Samuel J-L, Michel JB,Zongazo MA, de-la-Bastie D, Lompre A-M, Wisnewsky C, Rappaport L, Levy B, Schwartz K. Atrial natriuretic factor gene expression in rat ventricle during experimental hypertension. Am J Physiol 1989; 257:H979-H987.
27. Wåhlander H, Isgaard J, Jennische E, Friberg P. Left ventricular insulin-like growth factor I increases in early renal hypertension. Hypertension 1992; 19:25-32.
28. Russell-Jones DL, Leach RM, Ward JPT, Thomas CrR. Insulin-like growth factor-I gene expression is increased in the right ventricular hypertrophy induced by chronic hypoxia in the rat. J Mol Endocrinol 1993; 10:99-102.

29.Hanson MC, Fath KA, Alexander RW, Delafontaine P. Induction of cardiac insulin-like growth factor I gene expression in pressure overload hypertrophy. Am J Med Sci. 1993; 306(2):69-74.
30.Donohue TJ, Dworkin LD, Lango MN, Fliegner K, Lango RP, Benstein JA, Slater WR, Catanese VM. Induction of myocardial insulin-like growth factor-I gene expression in left ventricular hypertrophy. Circulation 1994; 89:799-809.
31.Isgaard J, Wahlander H, Adams MA, Friberg P Increased expression of growth hormone receptor mRNA and insulin-like growth factor-I mRNA in volume-overloaded hearts. Hypertension 1994; 23:884-888.
32.Toyozaki T, Hiroe M, Hasumi M, Horie T, Hosoda S, Tsushima T, Sekiguchi M. Insulin-like growth factor I receptors in human cardiac myocytes and their relation to myocardial hypertrophy. Jpn Circ J 1993; 57:1120-1127.
33.Donath MY, Gosteli-Peter MA, Hauri C, Froesch ER, Zapf J. Insulin-like growth factor I stimulate myofibrillar genes and modulates atrial natriuretic factor mRNA in rat heart. Eur J Endocrinol 1997; 137:309-315.
34.Donath MY, Zierhut W, Gosteli-Peter MA, Hauri C, Froesch ER, Zapf J. Effects of IGF-I on cardiac growth and expression of mRNAs coding for cardiac proteins after induction of heart hypertrophy in the rat. Eur J Endocrinol 1998; 139:109-117.
35.Vetter U, Kupferschmid C, Lang D, Benz S. Insulin-like growth factors and insulin increases the contractility of neonatal rat cardiocytes in vitro. Basic Research Cardiol 1988; 83:647-654.
36.Freestone NS; Ribaric S; Mason WT, The effect of insulin-like growth factor I on adult rat cardiac contractility. Mol Cell Biochem 1996; 163/164:223-229.
37.Ren J, Walsh MF, Hamaty M, Sowers JR, Brown RA. Altered inotropic response to IGF-I in diabetic rat heart: influence of intracellular Ca2+ and NO. Am J Physiol 1998; 275:H823-830.
38.Kinugawa S, Tsutsui H, Ide T, Nakamura R, Arimura K-I, Egashira K, Takeshita A. Positive inotropic effect of insulin-like growth factor-1 on normal and failing cardiac myocytes. Cardiovascular Research 1999; 43:157-164.
39.Cittadini A, Ishiguro Y, Stromer H, Spindler M, Moses AC, Clark R, Douglas PS, Ingwall JS, Morgan JP Insulin-like growth factor-1 but not growth hormone augments mammalian myocardial contractility by sensitizing the myofilament to Ca2+ through a wortmannin-sensitive pathway: studies in rat and ferret isolated muscles. Circ Res 1998; 83:50-59.
40.Ren J, Sowers JR, Natavio M, Brown RA. Influence of age on inotropic response to insulin and insulin-like growth factor I in spontaneously hypertensive rats: role of nitric oxide. Proc Soc Exp Biol Med 1999; 221:46-52.
41.Rodriguez-Tarduchy G, Collins MK, Garcia I, and Lopez-Rivas A. Insulin-like growth factor I inhibits apoptosis in IL-3-dependent hemopoietic cells. J Immunol 1992; 149:535-540.
42.Muta K, and Krantz SB. Apoptosis of human erythroid colony-forming cells is decreased by stem cell factor and insulin-like growth factor I as well as erythropoietin. J Cell Physiol 1993; 156:264-274.
43.Muta K, Krantz SB, Bondurant MC, Wickrema A. Distinct roles of erythropoietin, insulin-like growth factor I, and stem cell factor in the development of erythroid progenitor cells. J Clin Invest 1994; 94:34-43.
44.Gluckman BD, Guan J, Beilharz EJ, Klempt ND, Klempt M, Miller O, Sirimanne E, Dragunow M, Williams CE. The role of the insulin-like growth factor system in neuronal rescue. Ann NY Academy of Science 1993; 692:138:148.

45.Neff NT, Perevette D, Houenou LJ, Lewis ME, Glicksman MA, Yin QW, Oppenheim RW. Insulin-like growth factors: putative muscle-derived trophic agents that promote motoneuron survival. J Neurobiol 1993; 24:1578-1588.
46.Christofori G, Naik B, and Hanahan D. A second signal supplied by insulin-like growth factor II in oncogene-induced tumorigenesis. Nature 1994; 369:414-417.
47.Resnicoff M, Abraham D, Yutanawiboonchai W, Rotman HL, Kajstura J, Rubin R, Zoltick B, Baserga R. The insulin-like growth factor I receptor protects tumor cells from apoptosis in vivo. Cancer Res 1995; 55:2463-2469.
48.Wu X, Fan Z, Masui H, Rosen N, Mendelsohn J. Apoptosis induced by an anti-epidermal growth factor receptor monoclonal antibody in a human colorectal carcinoma cell line and its delay by insulin. J Clin. Invest 1995; 95:1897-1905.
49.Bennett MR, Evan GI, Schwartz SM. Apoptosis of human vascular smooth muscle cells derived from normal vessels and coronary arteriosclerotic plaques. J Clin Invest 1995; 95:2266-2274.
50.Buerke M, Murohara T, Skurk C, Nuss C, Tomaselli K, Lefer AM. Cardioprotective effect of insulin-like growth factor I in myocardial ischemia followed by reperfusion. Proc Natl Acad Sci USA 1995; 92:8031-8035.
51.Li Q, Li B, Wang X, Leri A, Jana KP, Liu Y, Kajstura J, Baserga R, Anversa P. Overexpression of insulin-like growth factor-1 in mice protects from myocyte death after infarction, attenuating ventricular dilation, wall stress, and cardiac hypertrophy. J Clin Invest 1997; 100:1991-1999.
52.Wang L, Ma W, Markovich R, Chen JW, Wang PH. Regulation of cardiomyocyte apoptotic signaling by insulin-like growth factor I. Circ Res 1998; 83:516-522.
53.Leri A, Liu Y, Claudio PP, Kajstura J, Wang X, Wang S, Kang P, Malhotra A, Anversa P. Insulin-like growth factor-1 induces Mdm2 and down-regulates p53, attenuating the myocyte renin-angiotensin system and stretch-mediated apoptosis. Am J Pathol 1999; 154:567-580.
54.Cittadini A, Stromer H, Katz SE, Clark R, Moses AC, Morgan JP, Douglas PS. Differential cardiac effects of growth hormone and insulin-like growth factor-1 in the rat. A combined in vivo and in vitro evaluation. Circulation. 1996; 93:800-809.
55.Duerr RL, Huang S, Miraliakbar HR, Clark R, Chien KR, Ross J. Insulin-like Growth factor-1 enhances ventricular hypertrophy and function during the onset of experimental cardiac failure. J Clin Invest 1995; 95:619-627.
56.Stromer H, Cittadini A, Douglas PS, Morgan JP. Exogenously administered growth hormone and insulin-like growth factor-I alter intracellular Ca2+ handling and enhance cardiac performance. In vitro evaluation in the isolated isovolumic buffer-perfused rat heart. Circ Res 1996; 79:227-236.
57.Ross J Jr, Hongo M. The role of hypertrophy and growth factors in heart failure. J Card Fail 1996; 2:S121-S128.
58.Ambler GR, Johnston BM, Maxwell L, Gavin JB, Gluckman PD. Improvement of doxorubicin induced cardiomyopathy in rats treated with insulin-like growth factor I. Cardiovasc Res 1993; 27:1368-1373.
59.Copland AC, Nair KS. Recombinant human insulin-like growth factor I increases forearm blood flow. J Clin Endocrinol Metab 1994; 79:230-232.
60.Fryburg DA. IGF I exerts GH and insulin-like actions on human muscle protein metabolism. Am J Physiol 1994; 267:E331-E336.
61.Kiowski W, Stoschitsky K, Kim JH, Hussain MA, Froesch ER. Insulin-like growth factor-I causes endothelium-dependent arterial vasodilatation. Circulation 1994; 90:I-242 (Abstr.).

62.Russell-Jones DL, Bates AT, Umpleby AM, Hennessy TR, Bowes SB, Hopkins KD, Jackson N, Kelly J, Shojaee-Moradie F, Jones RH, Sonksen PH. A comparison of the effects of IGF-I and insulin on glucose metabolism, fat metabolism and the cardiovascular system in normal human volunteers. Eur J Clin Invest 1995; 25:403-411.
63.Donath MY, Jenni R, Brunner HP, Anrig M, Kohli S, Glatz Y, Froesch ER. Cardiovascular and metabolic effects of insulin-like growth factor I at rest and during exercise in humans. J Clin Endocrinol Metab 1996; 81:4089-4094.
64.Bisi G, Podio V, Valetto MR, Broglio F, Bertuccio G, DEl Rio G, Boghen MF, Berti F, Muller EE, Ghigo E. Radionuclide angiocardiographic evaluation of the cardiovascular effects of recombinant human IGF-I in normal adults. Eur J Endocrinol 1999; 140:322-327.
65.Donath MY, Sutsch G, Yan XW, Piva B, Brunner HP, Glatz Y, Zapf J, Follath F, Froesch ER, Kiowski W. Acute cardiovascular effects of insulin-like growth factor I in patients with chronic heart failure. J Clin Endocrinol Metab 1998; 83:3177-3183.
66.McGuffin Jr WL, Sherman BM, Roth J, Gorden P, Kahn CR, Roberts WC, Frommer PL. Acromegaly and cardiovascular disorders. A prospective study. Ann Intern Med 1974; 81:11-18.
67.Martins JB, Kerber RE, Sherman BM, Markus ML, Ehrhardt JC. Cardiac function in acromegaly. Circulation 1977; 56:863-869.
68.Møller N, Schmitz O, Jørgensen JOL, Astrup J, Bak JF, Christensen SE, Alberti KG, Weeke J. Basal and insulin-stimulated substrate metabolism in patients with active acromegaly before and after adenomectomy. J Clin Endocrinol Metab 1992; 74:1012-1019.
69.Nikkila EA, Pelkonen R. Serum lipids in acromegaly. Metabolism 1975; 24:829-838.
70.Guler H-P, Schmid C, Zapf J, Froesch ER. Effects of recombinant insulin-like growth factor I on insulin secretion and renal function in normal human subjects. Proc Natl Acad Sci USA 1989; 86:2868-2872.
71.Zenobi PD, Graf S, Ursprung H, Froesch ER. Effects of insulin-like growth factor I on glucose tolerance, insulin levels and insulin secretion. J Clin Invest. 1992; 89:1908-1913.
72.Froesch ER, Hussain MA. Recombinant human insulin-like growth factor-I: a therapeutic challenge for diabetes mellitus. Diabetologia 1994; 37:179-185.
73.Zenobi PD, Holzmann P, Glatz Y, Riesen WE, Froesch ER. Improvement of lipid profile in type 2 (non-insulin-dependent) diabetes mellitus by insulin-like growth factor I. Diabetologia 1993; 36:465-469.
74.Bianda T, Zapf J, Froesch ER, Schmid C. Insulin-like growth factor-I lowers fasting and postprandial triglyceride levels without affecting chylomicron clearance in healthy men. J Intern Med 1999; 245:455-462.

19
GROWTH HORMONE-RELEASING PEPTIDES AND THE HEART

Vittorio Locatelli[1], Giuseppe Rossoni[2,3], Antonio Torsello[1], Micaela Bernareggi[1], Elena Bresciani[2], Vito De Gennaro Colonna[2], Romano Deghenghi[4], Eugenio E. Müller[2], Ferruccio Berti[1]

[1]School of Medicine, University of Milano-Bicocca; [2]Dept. of Pharmacology, [3]Inst. of Pharmacological Sciences, University of Milano, Italy; and [4]Europeptides, Argenteuil, France

INTRODUCTION

In the past few years, search of compounds alternative to either GH or GHRH, endowed with oral bioavailability and high effectiveness, has resulted in the development of a new family of molecules named growth hormone-releasing peptides (GHRPs). This family now comprises a number of different short peptidyl and non-peptidyl agonists, some of which have been already proven to be effective in humans, and for this reason these compounds are more generally referred to as the growth hormone secretagogues (GHSs). At present, no cardiac effects of the non-peptidyl compounds have been reported and therefore in this chapter we will focus on the cardiovascular actions of GHRPs.

IDENTIFICATION OF THE GROWTH HORMONE-RELEASING PEPTIDES

Many years before native GHRH had been isolated and sequenced (1, 2), the work by Bowers and associates (3) led to the synthesis of an enkephalin analog, devoid of any opioid activity, which was capable of specifically releasing GH from *in vitro* pituitaries (4). Though this pentapeptide, Tyr-DTrp-Gly-Phe-Met-NH_2, had a relatively weak potency as GH releaser and was active only *in vitro*, it served as a model for designing new molecules with much greater activity (5). Among them, GHRP-6 (His-DTrp-Ala-Trp-DPhe-Lys-NH_2) was considered particularly interesting because of its potent and specific GH-releasing effect exerted directly on the pituitary. The GH-releasing activity of GHRP-6 was demonstrated both *in vitro* and *in vivo* in a wide range of species, i.e monkeys, sheep, pigs, chicks, steers, rats (6-10), as well as man (11, 12). Moreover, GHRP-6 was active orally in rats, dogs, monkeys and humans (13, 14). In these studies, the stimulation of GH release appeared to be specific and without associated adverse effects, the only exception being mild sweating that occurred occasionally. In old dogs, the GHRP-6 analog hexarelin augmented the indices of spontaneous pulsatility of GH, reduced bone resorption, improved some biochemical and morphological muscular indices, though it did not induce changes of plasma IGF-I levels (15).

THE STRUCTURE-ACTIVITY RELATIONSHIPS

A very specific configuration of selected aromatic rings is required for all these peptides to specifically stimulate GH release. The pentapeptide, Tyr-DTrp-Gly-Phe-Met-NH_2, served as the starting model for the development of more potent GHRPs. As described by Momany et al (5), the *in vitro* and *in vivo* activity results primarily from the charged side chain of the N-terminal histidine and the central aromatic residues. GHRP-6 is the prototype starting from which many different analogs ranging 8 to 3 amino acid residues have been synthesized and shown to be active also in man (Tyr-Ala-hexarelin, GHRP-1, GHRP-2, hexarelin, ipamorelin etc.; Table 1) (16-20).

Recently, new peptides with reduced molecular size and complexity have been synthesized in the hope of understanding the minimal and tridimensional structure required for the GHRPs. Among them a pentapeptide (G-7039), a tetrapeptide (G-7134), a pseudotripeptide (G-7502), and a rigid heptapeptide (G-7203) were active at the putative GHRP receptor expressed in anterior pituitary cells (21). After 45 min of continuous exposure to these compounds, homologous desensitization developed, but the cells remained sensitive to GHRH (21). Another tetrapeptide, named tetrarelin (22), was shown to be active *in vivo*. Also cyclic analogs of GHRPs have been developed and shown to be very effective both *in vitro* and *in vivo* (23).

Although GHRPs were initially introduced as specific GH-secretagogues, small rises in serum cortisol and prolactin (PRL) occurring in humans after iv infusion, bolus injection or intranasal administration of peptidyl GHSs were disclosed (12). Stimulation of GH and PRL secretion from rat primary pituitary cell cultures has also been reported (23). The real impact of these observations in situations of long-term GHS therapy is presently unknown. Despite the potential of affecting the secretion of more than one anterior pituitary hormone, nowadays GHSs are the most effective GH-secretagogues known and could be profitably used in humans with GH hyposecretory disturbances to promote a pattern of GH secretion that mimics physiology in a better way than the exogenously administered GH.

Table 1. The structure of representative GHRPs.

Name	Structure
GHRP-6	His-D-Trp-Ala-Trp-D-Phe-Lys-NH_2
Hexarelin	His-D-Trp(2-Me)-Ala-Trp-D-Phe-Lys-NH_2
GHRP-2	D-Ala-D-ßNal-Ala-Trp-D-Phe-Lys-NH_2
Ipamorelin	Aib-His-D-2-Nal-D-Phe-Lys-NH_2
G-7039	Inip-D-ßNal- D-ßNal-Phe-Lys-NH_2
Tetrarelin	Gab-D-Trp(2-Me)-D-ßNal-Phe-Lys-NH_2

Aib=Aminoisobutyryc acid; Inip = isonipecotic acid; ßNal = ß-(2-naphthyl)alanine; Gab = gamma-aminobutyryl

The strategy of stimulating GH secretion by activating the endogenous pituitary machinery is an attractive alternative that has also been pursued with the synthesis of non-peptidyl mimics of GHRP-6, a research that has led to the discovery of a novel series of compounds, the benzolactam secretagogues (24). A lead compound, the biphenylcarboxylic acid derivative (L-158,077), identified through direct screening of non-peptidyl templates, stimulated GH release from rat primary pituitary cultures in a dose-dependent manner. Further modifications of the molecule gave L-692,429, which proved to be the first well tolerated, non-peptidyl GH secretagogue in animal and man, and the spiropiperidine L-163,191 (MK-677) that is characterized by high oral bioavailability. These new developed molecules are very effective in eliciting GH release, even if they also stimulate *in vivo* the release of ACTH and PRL and MK-677 also increases fasting blood glucose and insulin concentrations (25-27). Although their oral bioactivity is much greater than that of the peptidic secretagogues, clinical use of these non-peptidyl molecules is currently hampered by a non satisfying understanding of the pharmacodynamic profile required for optimal efficacy.

MECHANISM OF ACTION IN THE PITUITARY

It has been clearly demonstrated that GHRH and GHSs act on the pituitary through different mechanisms and via distinct receptors. Somatotrophs that are maximally stimulated with GHRPs can release additional GH in response to GHRH and viceversa (28, 29). Moreover, continuous exposure of the pituitary to GHSs or GHRH induced homologous, but not heterologous, desensitization (28, 30-32). At the beginning of the studies with GHSs it was thought that GHRP-6 was acting on the GHRH receptor because it failed to elicit GH release in the lit/lit mouse, an animal model with a point mutation in the GHRH receptor (33). However, against this view data obtained in cultures of human somatotropinomas evidenced a GH responsiveness to GHRP-6 in all adenomas, whereas only half of them responded to GHRH (34). As an alternative possibility, Elias et al (21) have suggested that despite the presence of mutually exclusive binding sites for GHRH and GHSs, in the pituitary the postreceptor activity triggered by GHRH may be instrumental to GHSs action. In contrast to the marked synergy observed *in vivo*, only additive effects are evident when GHSs and GHRH are coincubated *in vitro* (28, 29, 35).

It is widely recognized that the GH release induced by GHRH is paralleled by an increase in cAMP titers in the pituitary, followed by the opening of voltage-dependent calcium channels (VDCCs). The rapid increase in intracellular calcium concentrations promotes the release of GH. In contrast, GHRP-6 alone is devoid of such effects on cAMP (30, 32), but a synergistic increase of cAMP levels occurs when GHRP-6 and GHRH are coadministered (30).

Although the precise definition of the molecular mechanism(s) subserving the action of GHSs is still lacking, these compounds elicit an increase in intracellular calcium derived by an extracellular source, since it can be blocked by the chelation of extracellular calcium or incubation with calcium channel blockers (36, 37). Accordingly, it has also been shown that GHRP-1 and GHRP-6 induce depolarization of rat pituitary cell membranes, leading to opening of VDCCs (38), and it has been proposed that GHRP-6 and L-692,429 may act, at least in part, through activation of the protein kinase C pathway (39, 40).

More recently, Pong and coll (41) have demonstrated in porcine and rat pituitary membranes a specific binding site for GHSs, distinct from that for GHRH and with the properties of a new G-protein-coupled receptor, and Van der Ploeg and associates have reported the cloning of a receptor for GHRP-6 and non-peptidyl GHS (42). Studies on pituitary and hypothalamic membrane preparations have revealed the existence of multiple binding sites with different affinity for GHSs and non peptidyl compounds. In the rat pituitary, MK-677 and hexarelin would bind preferentially to an high affinity site (Kd in the pM range), whereas medium (nM range) and low affinity (μM range) sites would primarily bind GHRP-6 (41-44). In pig pituitaries only the high and low affinity sites were expressed (41, 45). Similar results were also obtained in hypothalamic membrane preparations.

In all, abundant experimental evidence indicates that peptidyl and nonpeptidyl GHS analogues operate through common cellular mechanisms, which involve the activation of a receptor(s) which is not that of GHRH, and the activation of intracellular mechanisms different from the cAMP pathways. Although the GHS receptor (or, most likely, one of its subtypes) has been cloned in man, rat and swine (42, 46), we presently ignore the existence and the nature of the purported endogenous GHS ligand. More recently, evidence for GHS receptor subtypes in rat pituitary and heart, distinct from that previously cloned, was obtained using a photoactivable analog of hexarelin (47, 48).

CARDIAC ACTIVITY OF GHRPs: EXPERIMENTAL MODELS

Over the past few years, significant advances have been made in our understanding of the cellular and molecular mechanisms involved in growth hormone (GH) action, including its effects on cardiac tissue (Chapter 1). Furthermore, experimental evidence points to a role of GH in cardiac pathophysiology, as discussed with more details in Chapters 3-7. We have recently obtained experimental evidence that hexarelin, administered in vivo in the rat, may show important cardioprotective effects, in particular against the ischemia-reperfusion damage. The experimental model consisted of isolated hearts that were perfused retrogradly through the aorta with gassed buffer at a flow rate of 12-15 ml/min (49). The preparations were electrically paced at a frequency of 300 beats/min with rectangular pulses. Ischemia was induced by reducing the coronary flow to 2 ml/min for a period of 20-40 min. At the end of this period, reperfusion was resumed at the pre-ischemic flow rate. Since hexarelin is a very potent and effective GH releaser, we wanted to ascertain whether hexarelin had direct effects on the heart or its cardioprotective actions were mediated by GH release. With these aims, experiments were performed in normal rats and experimental models of GH deficiency (rats passively immunized against GHRH, aged rats with spontaneous reduction of GH release and hypophysectomized rats).

Young-Adult Rats with Isolated GH Deficiency

Recently, we have reported that heart preparations from a rat model of selective GH deficiency (GHD), when subjected to global flow reduction and reperfusion, exhibited clear signs of cardiac dysfunction, consisting of ventricular contracture during ischemia and severe impairment of strenght of contractility at reperfusion. In this instance coronary resistance was significantly increased. In these rats, GHD was induced by passive immunization with polyclonal antibodies against GHRH for 3 weeks. Interestingly, GH replacement therapy prevented ischemic and postischemic myocardial abnormalities (50).

In heart preparations from normal rats, reduction of perfusion flow resulted in minor ischemic damage at reperfusion (Figures 1 and 2). Based on a prompt restoration of electrical pacing and on a minimal increase in resting tension, left ventricular developed pressure (LVDP) recovered to 58% of basal values within 20 min. In hearts from GHD rats, LVDP recovered only 30% of basal values. Moreover, the values of left ventricular end-diastolic pressure (LVEDP) at the end of the ischemic and reperfusion periods were almost 5 to 8-fold higher than the corresponding values in normal rats (Figure 1). Rhythm disturbances were present and coronary perfusion pressure (CPP) was increasedby 52% compared to normal rats.

Hexarelin treatment (80 μg/kg, sc, bid) for 15 days reduced ischemic damage in the heart of GHD rats. In fact, at the end of ischemic and reperfusion periods, LVEDP was only 1.4-fold and 2.3-fold higher than in preparations from normal rats treated with saline (Figure 1). At the end of the reperfusion period, CPP was in the normal range (Figure 2). The vasopressor activity of angiotensin II (AII), 1 μg/bolus, on CPP was monitored at the beginning of each experiments. In GHD rats, AII bolus induced a CPP rise 3.8-fold higher than that measured in heart preparations from normal rats (Figure 3). In contrast, this hyper-reactivity to AII was not seen in hearts from GHD rats treated with hexarelin (Figure 3).

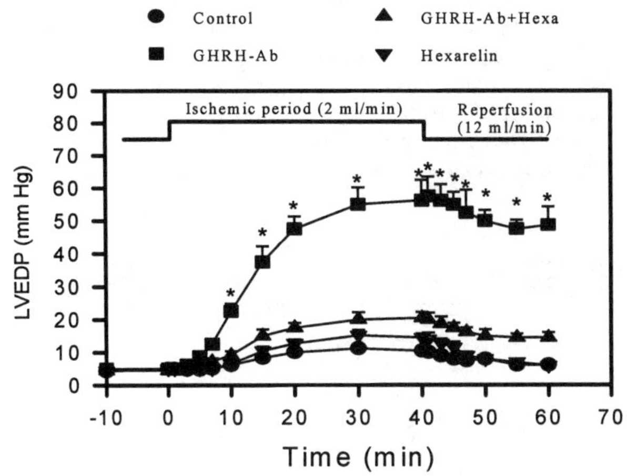

Figure 1. Effects of low-flow ischemia and reperfusion on left ventricular end-diastolic pressure (LVEDP). Rats were treated *in vivo* with polyclonal antibodies against GHRH (GHRH-Ab) or preimmune rabbit serum (control) for 3 weeks. Two groups were also given hexarelin or physiological saline for 2 weeks. At the end of treatment, paced isovolumic left heart preparations were exposed to low-flow ischemia and reperfusion. Values are the means ± SEM from ten hearts. * P<0.01 vs all other groups.

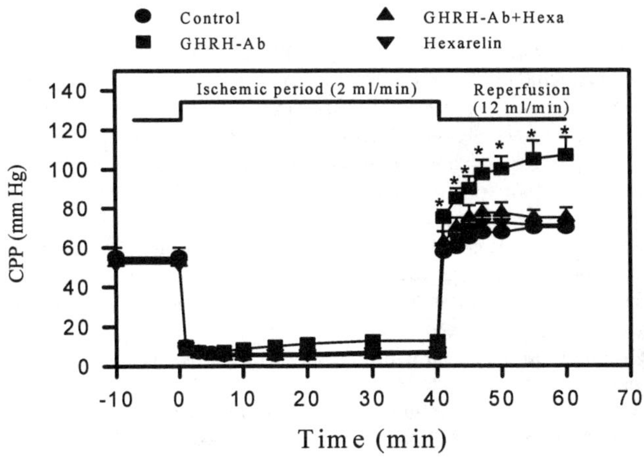

Figure 2. Effects of low-flow ischemia and reperfusion on coronary perfusion pressure (CPP). Same heart preparations of Figure 1. Values are the means ± SEM from ten hearts. * P<0.01 vs all other groups.

Hexarelin treatment had significant positive effects on the cardiac parameters also in normal rats. These results demonstrate that the administration of hexarelin to rats with selective GHD is capable of preventing the post-ischemic ventricular dysfunction to an extent similar to GH. Moreover, the observation that hexarelin induced a clear-cut protection against myocardial damage also in control rats raises the important issue of its mechanism of action on the heart and suggest that the peptide may act directly on the heart (51). However, direct perfusion of hexarelin through the heart had no effects, indicating that hexarelin actions require some still unknown modifications of the cardiac myocyte physiology (unpublished data).

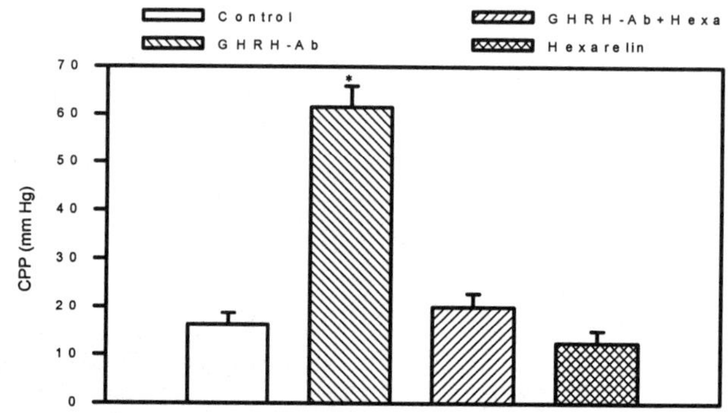

Figure 3. Vasopressor activity of angiotensin II injected into isovolumic left heart preparation during the pre-ischemic phase. Same heart preparations of Figure 1. Values are the means ± SEM of 10 hearts. * $P<0.01$ vs all other groups.

Aged Rats

It has been shown that aging alters the spectrum of physiological and biochemical properties of the myocardium, including force production, excitation-contraction coupling, substrate use and mitochondrial oxidative capacity (52). New insights into myocardial-reperfusion injury indicate that aged rats, besides a reduction of the myocardial antioxidant defense mechanisms (53), suffer from alterations of calcium handling in cardiac cells (54). In fact, abnormalities of regulation/modulation mechanisms normally involved in the restriction of calcium oscillation between sarcoplasmic reticulum and cytoplasm are associated with strong impairment of the cardiac mechanics. These findings and the awareness that GH secretion and its biological effects decline with aging in both experimental animals and humans (55-56) prompted us to investigate the protective actions of hexarelin in this spontaneous model of GHD.

Treatment of 24-month-old male rats with hexarelin (80 µg/kg, sc, bid) or GH (400 µg/kg, sc, bid) for 21 days did not affect pituitary GH mRNA and plasma IGF-I levels. The global flow reduction for 20 min (from 15 ml/min to 1 ml/min) in isovolumic left heart preparations obtained from saline-treated rats induced a clear cut decrease of left ventricular function, associated with a substantial increase in coronary resistance. In fact, 30 min after the beginning of reperfusion the postischemic LVDP recovered only the 37% of preischemic values (Figure 4). At this time, CPP was still 71% over the basal values (Figure 5) and was not associated to stiffness of the hearts.

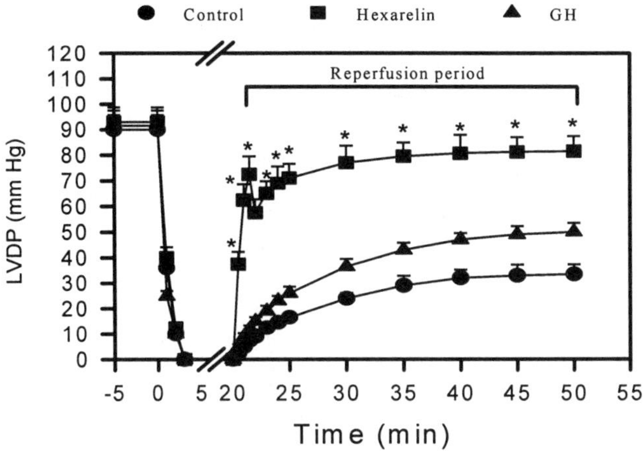

Figure 4. Effects of low-flow ischemia and reperfusion on LVDP in aged rats. Rats were treated *in vivo* with hexarelin or GH for 21 days. At the end of treatment, paced isovolumic left heart preparations were exposed to low-flow ischemia and reperfusion. Values are the means ± SEM from 6 hearts. * P<0.01 vs all other groups.

In heart preparations obtained from rats treated with hexarelin, a regular paced rhythm appeared at the beginning of reperfusion and 30 min later the recovery of postischemic LVDP was about 90% of preischemic values (Figure 4). In these hearts CPP values increased only marginally in the first 5 min of reperfusion, and basal values were attained by 30 min (Figure 5). A lower protective activity against reperfusion damage was present in hearts obtained from GH-treated rats. In these preparations, the trend of postischemic left ventricular dysfunction was similar to that present in saline-treated rats (Figures 4 and 5). Determination of creatine kinase (CK) activity in the heart perfusate is often assumed as a biochemical marker of myocardial cell lesions. In keeping with the clear cut decrease of left ventricular function, in the reperfusion period CK released in the effluent of hearts obtained from rats treated with saline or GH was significantly higher than in the preischemic phase (Figure 6). In contrast, in the hexarelin group CK increased only moderately and gradually returned to baseline by the end of reperfusion (Figure 6). Bolus injections of increasing doses of AII (0.25-4 μg) into heart preparations at the beginning of each experiment induced a dose-related increase in CPP. The vasopressor activity of AII was not statistically different in the three experimental groups, implying that in aged rats either hexarelin or GH did not interfere with the endothelium-dependent relaxant function of coronary vasculature.

Myocardial ischemia may have several potential outcomes, especially in senescent hearts that are more prone to this pathologic event. In our model of ischemia-reperfusion, hearts from old rats treated for 21 days with hexarelin achieved a strong protection against myocardial stunning. In fact, complete recovery of left ventricular function was present on reperfusion. Simultaneous blunting of CK leakage in the heart effluent bespoke the integrity of myocardial cell membranes and the preservation from the damage that follows oxygen readmission.

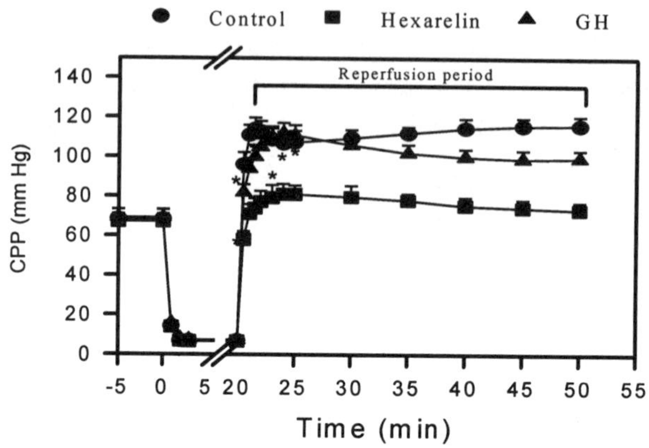

Figure 5. Effects of low-flow ischemia and reperfusion on CPP in aged rats. Same heart preparations of Figure 4. Values are the means ± SEM from 6 hearts. * P<0.01 vs all other groups.

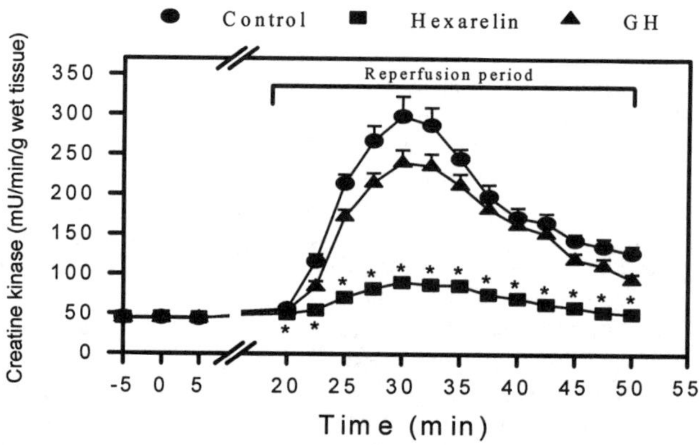

Figure 6. Creatine kinase release profile in ischemic and reperfusion conditions in aged rats. Same heart preparations of Figure 4. Values are the means ± SEM from 6 hearts. * P<0.01 vs all other groups.

An interesting feature is the weak protective effect exerted by GH treatment in these senescent rats. The improvement of postischemic ventricular function was, in fact, modest and by no means comparable with that elicited by hexarelin. GH action is very likely attributable to a direct action on its cardiac receptor. Another very important finding is the inability of GH and hexarelin to alter the vasopressor activity of AII: the dose-response curves of AII on CPP during the preischemic period were similar to those of the group treated with saline. In contrast to the data obtained in young-adult GHD rats, where the impairment of endothelial-dependent relaxant function was counteracted by the two treatments, our results indicate that GH and hexarelin failed to improve the function of coronary vascular endothelium in senescent rats.

Hypophysectomized Rats

At this point we have shown that hexarelin is a highly effective GHRP and proposed that its cardiac effects are likely mediated by GH (51). However, previous studies indicated that the GH secreting activity of hexarelin is largely impaired in the GHD rat model (57). Alternatively, hexarelin activity in the heart may be only partially dependent on GH or even independent of GH. If GH-independent, a possibility may be that hexarelin stimulates specific binding sites for GHS-like compounds in the heart (47, 58). To address this issue, we compared the effects of hexarelin to those of GH on the mechanical and metabolic alterations induced by low-flow ischemia and reperfusion in isolated hearts obtained from hypophysectomized rats treated for 1 week with hexarelin, GH or saline. To study the specificity of hexarelin on cardiac function we compared its effects with those of EP 51389, a synthetic tripeptide with strong GH-releasing activity (58, 59). The structure of EP 51389 is distinct from that of hexarelin and, therefore, it is unable to displace hexarelin from its cardiac binding sites (58). In the hypophysectomized rats, hexarelin or EP 51389 can not stimulate the release of pituitary GH, therefore the cardiac effects of these peptides must be GH-independent.

When ischemia-reperfusion was induced in hearts from saline-injected hypophysectomized rats, a marked ischemic damage occurred. In this instance, during the ischemic phase the values of LVEDP gradually increased and at the end of reperfusion, remained still elevated (Figure 7).

As a consequence, electrical pacing was not re-established and cardiac rhythm disturbances were associated with a severe impairment of heart contractility. Moreover, upon reperfusion, CPP values were significantly increased, denoting severe coronary vasoconstriction due, in part, to heart stiffness (Figure 8).

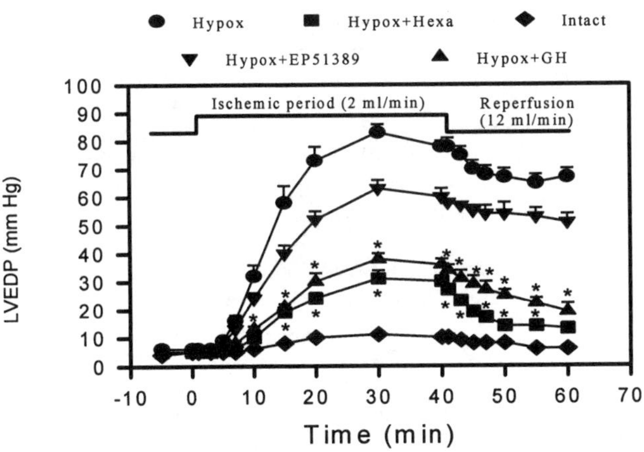

Figure 7. Effects of low-flow ischemia and reperfusion on LVEDP in hypophysectomized rats. Rats were treated *in vivo* with hexarelin or GH for 1 week. Values are the means ± SEM from ten hearts. * $P<0.01$ vs all other groups.

Treatment of hypophysectomized rats with hexarelin (80 μg/kg/day) notably protected the isolated hearts from ischemia-reperfusion damage such that CPP values were in the range of those determined in preparations from intact rats (Figure 8). At the end of the ischemic and reperfusion periods, LVEDP and CPP values of hypophysectomized rats treated with hexarelin were not statistically different from those of intact rats (Figure 7).

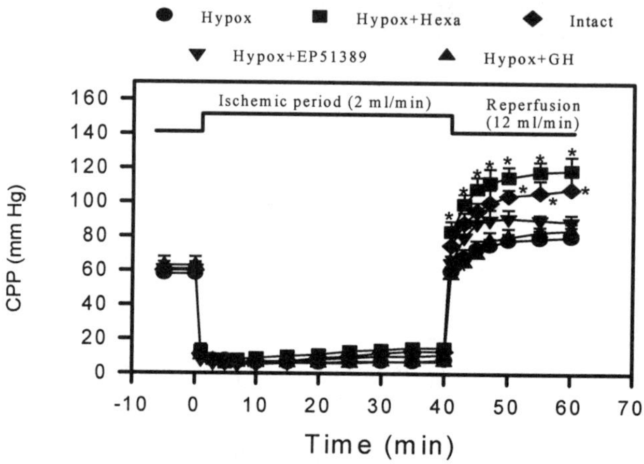

Figure 8. Effects of low-flow ischemia and reperfusion on CPP in hypophysectomized rats. Same heart preparations of Figure 7. Values are the the means ± SEM from 6 hearts. * P<0.01 vs all other groups.

Similar results were obtained with heart preparations from hypophysectomized rats given GH. In this instance, LVEDP and CPP (Figures 7 and 8) values were not statistically different from those obtained from hearts of hypophysectomized rats given hexarelin. In contrast, heart preparations from hypophysectomized rats given EP 51389 generated LVEDP values, both during ischemia and at the end of reperfusion, that were consistently greater than the corresponding pre-ischemic values (Figure 7). Dysrhythmia was present during reperfusion. At the end of the reperfusion period, these hearts recovered only 35% of their pre-ischemic contractility values. In line with these results, at the end of reperfusion the CPP values were still higher than basal values (Figure 8).

The CK activities found in heart perfusates collected during the reperfusion period exhibited a strong correlation with the degree of myocardial ischemic injury present in the five experimental groups. The total amount of CK released during the reperfusion period from hearts of hypophysectomized animals was almost 3-fold higher than that found in perfusates of intact rats (Figure 9).

Treatment with GH or hexarelin reduced, by almost 50%, the amount of CK released by the hearts of hypophysectomized rats during reperfusion. In contrast, heart preparations from hypophysectomized rats given EP 51389 generated CK activity in amounts similar to those released by hearts from saline-injected hypophysectomized rats.

Consistent with those data indicating that GH is needed to maintain optimal heart contractility, our results show that heart global flow limitation and reperfusion induced significantly greater myocardial damage in hypophysectomized rats than in intact animals. Compared to the latter, hearts from hypophysectomized rats presented with severe signs of ischemic and postischemic ventricular dysfunction, arrhythmia, increased CK activity in the perfusates, and constriction of the coronary vascular bed.

GH replacement reduced ischemic cardiac injury. Hearts from GH-treated hypophysectomized rats exhibited a quicker recovery of contractility than hearts from the saline group, and more promptly followed the imposed electrical pacing. This treatment also normalized CK activity in the perfusates and the vasoconstriction of the coronary vasculature induced by angiotensin II. GH may have acted directly on the myocardium stimulating specific receptors and/or indirectly, through an increase in circulating IGF-I levels.

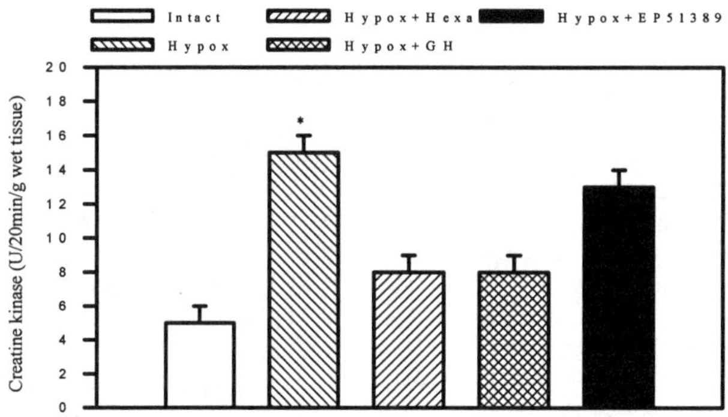

Figure 9. Creatine kinase release in reperfusion period in hypophysectomized rats. Same heart preparations of Figure 7. Values are the means ± SEM from 6 hearts. * P<0.01 vs all other groups.

In this study, hexarelin had a strong protective activity against ischemia and reperfusion-induced myocardial damage, very similar to that observed for GH. Hexarelin pretreatment effectively reduced the ventricular contracture of the perfused heart during ischemia, and reduced CK activity in the heart outflow at reperfusion. These events were paralleled by a more efficient recovery of LVDP, a prompt compliance of the heart to follow the external electrical pacing and a reduction of the coronary perfusion pressure. The protective effect of hexarelin was also demonstrated by restoration of the coronary vessel reactivity to angiotensin II.

The mechanism by which hexarelin exerts its beneficial effects on cardiac function in hypophysectomized rats is obviously independent of GH. Data obtained with the tripeptide EP 51389 are consistent with this view. This molecule is as effective as hexarelin in stimulating GH secretion in the rat (58, 59), but it is far less effective in protecting the heart from ischemia. Interestingly, EP 51389 effectively displaces hexarelin from its hypothalamic binding sites, and poorly from cardiac membranes (58), which suggests the presence of multiple receptor subtypes for GHS.

Another interesting finding is that the aortic rings prepared from hypophysectomized rats exhibited an increased sensitivity to norepinephrine and endothelin-1. These data indicate that, in hypophysectomized rats, a deficiency of the endothelium-dependent relaxant mechanism may be a widespread phenomenon involving different circulatory domains.

Treatment of these rats with GH or with hexarelin prevented almost entirely the functional abnormalities evidenced in the isolated aortic rings. In fact, in the vascular tissue of these rats, the responses to endothelin-1, acetylcholine or L-NMMA were very similar to those of intact rats. These data suggest that GHRPs, as hexarelin, may also have therapeutic value in delaying the initiation and progression of vascular disorders due to endothelial dysfunction.

In Vitro Studies

Very little data are available on the direct effects of hexarelin on cardiomyocytes. Interestingly, it has been very recently shown that hexarelin binds to specific high affinity binding sites on H9c2 cells, a fetal cardiomyocytes-derived cell line. In vitro, hexarelin promotes survival and inhibits apoptosis of H9c2 cells treated with cytotoxic agents (e.g. TNFα and doxorubicin). Since in this model hexarelin activates Akt serine kinase, it has been proposed that hexarelin is a survival factor for cardiomyocyte-derived cells, and that it exerts its antiapoptotic activity through a specific, receptor-mediated mechanism (60).

CARDIAC ACTIVITY OF GHRPs: HUMAN STUDIES

There is a dramatic paucity of data on the cardiac effects of GHRPs in humans. In one study, the cardiac performance was studied by radionuclide angiocardiography after the acute iv administration of hexarelin in 7 adult male GHD patients and in 7 adult male control subjects (61). The GH response to hexarelin was negligibile in GHD compared to normal controls. At baseline, left ventricular ejection fraction (LVEF) in GHD was lower than in normal controls. Hexarelin administration increased LVEF both in GHD and normal controls. Moreover, it enhanced stroke volume in both groups, but did not increase significantly cardiac outputs, likely due to a concomitant trend toward a decrease in heart rate. These data suggest that, in contrast to rats, in humans hexarelin exerts an acute, GH independent, positive inotropic effect.

CONCLUSIONS

We still ignore what kind of intracellular signal transduction is triggered by GHS-receptor activation in peripheral organs, a point that deserves a thorough investigation. However, the striking hexarelin-induced inhibition of reperfusion damage in the isolated hearts would call for a restraint in the increase of cytosolic calcium that follows reperfusion (62, 63). Such effect is reminiscent of that of calcium channel blockers, e.g. diltiazem and verapamil. However, since hexarelin perfused through the heart does not depress myocardial contractility, its mechanism of action is likely different from that of these drugs. In fact, it is known that calcium entry blockers protect the heart from the damage induced by ischemia-reperfusion by a strong depression of myocardial contractility ultimately related to inhibition of calcium entry into the myocardial cells (64, 65). It is then possible that sustained administration of hexarelin may have increased cardiomyocyte energy stores to an extent compatible with the maintenance of basic cell organization which allows a normal recovery of the contractile function at the end of ischemic insult. Our studies, however, still lack data on both the concentration of energy storing nucleotides and on the grade of density of glycogen granules in ischemic hearts of control and hexarelin treated-rats. These aspects, together with the evaluation of ultrastructural changes of myocardial cells associated with ischemia, may provide a clearer picture of the mode of action of hexarelin in postischemic ventricular dysfunction.

REFERENCES

1. Guillemin R, Brazeau P, Bohlen P, Esch F, Ling N, Wehrenberg WB. Growth hormone releasing factor from a human pancreatic tumor that caused acromegaly. Science 1982;218:585-587.
2. Rivier J, Spiess J, Thorner MO, Vale WW. Characterization of a growth hormone-releasing factor from a human pancreatic islet cell tumour. Nature 1982;300:276-278.
3. Bowers CY, Chang J, Momany F, Folkers K. Effect of the enkephalins and enkephalin analogs on release of pituitary hormones, in vitro. In: MacIntyre G, Szelke H eds. Molecular Endocrinology. Amsterdam: Elsevier/North Holland 1977:287-92.
4. Bowers CY, Momany F, Reynolds GA, Chang D, Hong A, Chang K. Structure-activity relationships of a synthetic pentapeptide that specifically releases growth hormone in vitro. Endocrinology 1980;106:663-667.
5. Momany FA, Bowers CY, Reynolds GA, Hong A, Newlander K. Conformational energy studies and in vitro and in vivo activity data on growth hormone-releasing peptides. Endocrinology 1984;114:1531-1536.
6. Bowers CY, Momany FA, Reynolds GA, Hong A: On the in vitro and in vivo activity of a new synthetic hexapeptide that acts on the pituitary to specifically release growth hormone. Endocrinology 1984;114:1537-1545.
7. Croom WJ, Leonard ES, Baker PK, Kraft LA, Ricks CA. The effects of synthetic growth hormone releasing hexapeptide BI 679 on serum growth hormone levels and production in lactating dairy cattle. J Anim Sci 1984;67:109.
8. Kraft LA, Baker PK, Doscher ME, Ricks CA. Effect of a synthetic growth hormone releasing hexapeptide on serum growth hormone levels in steers. J Anim Sci 1984;59:218.
9. Doscher ME, Baker PK, Kraft LA, Ricks CA. Effect of a synthetic growth hormone releasing hexapeptide (BI 679) and growth hormone releasing factor (GRF) on serum growth hormone levels in barrows. J Anim Sci 1984;59:218.
10. Malozowski S, Hao EN, Ren SG, et al. Growth hormone (GH) responses to the hexapeptide GH-releasing peptide and GH-releasing hormone (GHRH) in the cynomologus macaque: evidence for non-GHRH-mediated responses. J Clin Endocrinol Metab 1991;73:314-317.
11. Ilson BE, Jorkasky DK, Curnow RT, Stote RM. Effect of a new synthetic hexapeptide to selectively stimulate growth hormone release in healthy human subjects. J Clin Endocrinol Metab 1989;69:212-4.
12. Bowers CY, Reynolds GA, Durham D, Barrera CM, Pezzoli SS, Thorner MO. Growth hormone (GH)-releasing peptide stimulates GH release in normal men and acts synergistically with GH-releasing hormone. J Clin Endocrinol Metab 1990;70:975-982.
13. Walker RF, Codd EE, Barone FC, Nelson AH, Goodwin T, Campbell SA. Oral activity of the growth hormone releasing peptide His-DTrp-Ala-Trp-DPhe-Lys-NH2 in rat, dogs, monkeys. Life Sci 1990;47:29-36.
14. Bowers CY, Alster DK, Frentz JM. The growth hormone-releasing activity of a synthetic hexapeptide in normal men and short statured children after oral administration. J Clin Endocrinol Metab 1992; 74:292-298.
15. Cella SG, Cerri CG, Daniel S, et al. Sixteen weeks of Hexarelin therapy in aged dogs: effecs on the somatotropic axis, muscle morphology, and bone metabolism. J Gerontol 1996;51A:B439-B447.

16. Arvat E, Di Vito L, Lanfranco F, et al. Tyr-Ala-Hexarelin, a synthetic octapeptide, possesses the same endocrine activities of Hexarelin and GHRP-2 in humans. J Endocrinol Invest 1999;22:91-97.
17. Deghenghi R, Cananzi MM, Torsello A, Battisti C, Muller EE, Locatelli V. GH-releasing activity of Hexarelin, a new growth hormone-releasing peptide, in infant and adult rats. Life Sci 1994;54:1321-1328.
18. Raun K, Hansen BS, Johansen NL, et al. Ipamorelin, the first selective growth hormone secretagogue. Eur J Endocrinol 1998;39:552-561.
19. Akman MS, Girard M, O'Brien LF, Ho AK, Chik CL. Mechanisms of action of asecond generation growth hormone-releasing peptide (Ala-His-D-beta Nal-Ala-Trp-D-Phe-Lys-NH2) in rat anterior pituitary cells Endocrinology 1993;132:1286-1291.
20. McCormick GF, Millard WJ, Badger TM, Bowers CY, Martin JB. Dose-response characteristics of various peptides with growth hormone-releasing activity in the unanesthetized male rat. Endocrinology 1985;117:97-105.
21. Elias KA, Ingle GS, Burnier JP, et al. In vitro characterization of four novel classes of growth hormone-releasing peptide. Endocrinology 1995;136:5694-5699.
22. Deghenghi R, Boutignon F, Luoni M, Grilli R, Guidi M, Locatelli V. Small peptides as potent releasers of growth hormone. J Ped Endocrin Met 1995;8:311-313.
23. McDowell RS, Elias KA, Stanley MS, et al. Growth hormone secretagogues: Characterization, efficacy, and minimal bioactive conformation. Proc Natl Acad Sci USA 1995;92:11165-11169.
24. Smith RG, Cheng K, Schoen WR, et al. A nonpeptidyl growth hormone secretagogue. Science 1993;260:1640-1643.
25. Gertz BJ, Barrett JS, Eisenhandler R, et al. Growth hormone response in man to L-692,429, a novel nonpeptide mimic of growth hormone-releasing peptide-6. J Clin Endocrinol Metab 1993;77:1393-1397.
26 Hickey G, Jacks T, Judith F, et al. Efficacy and specificity of L-692,429, a novel nonpeptidyl growth hormone secretagogue, in beagles. Endocrinology 1994;134:695-701.
27. Chapman IM, Bach MA, Van Cauter E, et al. Stimulation of the growth hormone (GH)-insulin-like growth factor I axis by daily oral administration of a GH secretagogue (MK-677) in healthy elderly subjects. J Clin Endocrinol Metab 1996;81:4249-4257.
28. Blake AD, Smith RG. Desensitization studies using perifused rat pituitary cells show that growth hormone-releasing hormone and His-D-Trp-D-Phe-Lys-NH2 stimulate GH release through different receptor sites. J Endocrinol 1991;129:11-19.
29. Cheng K, Chan WW, Barreto A, Convey EM, Smith RG. The synergistic effects of His-D-Trp-Ala-Trp-D-Phe-Lys-NH2 on growth hormone (GH)-releasing factor-stimulated GH release and intracellular adenosine 3',5'-monophosphate accumulation in rat primary pituitary cell culture. Endocrinology 1989;124:2791-2798.
30. Clark RG, Carlsson LMS, Trojnar J, Robinson ICAF. The effects of a growth hormone-releasing peptide and growth hormone-releasing factor in conscious and anaesthetized rats. J Neuroendocrinology 1989:1;249-255.
31. Chang CH, Rickes EL, Marsilio F, et al. Activity of a novel nonpeptidyl growth hormone secretagogue, L-700,653, in swine. Endocrinology 1995;136:1065-1071.
32. Wu D, Chen C, Katoh K, Zhang J, Clarke J. The effect of GH-releasing peptide-2 (GHRP-2 or KP 102) on GH secretion from cultured ovine pituitary cells can be abolished by a specific GH-releasing factor (GRF) receptor antagonist. J Endocrinol 1994:140;9-13.

33.Jansson J-O, Downs TR, Beamer WG, Frohman LA. The dwarf "little" (lit/lit) mouse is resistant to growth hormone (GH)-releasing peptide (GH-RP-6) as well as to GH-releasing hormone (GRH). 68th Ann Meeting Endocr Soc 1986 abs 397.
34.Renner U, Brockmeier S, Strasburger CJ, et al. Growth hormone (GH)-releasing peptide stimulation of GH release from human somatotroph adenoma cells: interaction with GH-releasing hormone, thyrotropin-releasing hormone, and octreotide. J Clin Endocrinol Metab 1994:78;1090-1096.
35.Badger TM, Millard WJ, McCormick GF, Bowers CY, Martin JB. The effects of growth hormone (GH)-releasing peptides on GH secretion in perifused pituitary cells of adult male rats. Endocrinology 1984:115;1432-1438.
36.Akman MS, Girard M, O'Brien LF, Ho AK, Chik CL. Mechanisms of action of a second generation growth hormone-releasing peptide (Ala-His-D-beta-Nal-Ala-Trp-D-Phe-Lys-NH2) in rat anterior pituitary cells. Endocrinology 1993:132;1286-1291.
37.Sartor O, Bowers CY, Chang D. Parallel studies of His-D-Trp-Ala-Trp-D-Phe-Lys-NH2 in rat primary pituitary cell monolayer culture. Endocrinology 1985;116:952-957.
38.Pong SS, Chaung L-YP, Smith RG. GHRP-6 (His-D-Trp-Ala-Trp-D-Phe-Lys-NH2) stimulates growth hormone secretion by depolarization in rat pituitary cell cultures. 73th Ann Meeting Endocr Soc 1991 abs 88.
39.Cheng K, Chan WW, Butler B, Barreto A, Smith RG. Evidence for a role of protein kinase-C in His-D-Trp-D-Phe-Lys-NH2-induced growth hormone release from rat primary pituitray cells. Endocrinology 1991;129:3337-3342.
40.Cheng K, Chan WW, Butler B, et al. Stimulation of growth hormone release from rat pituitary cells by L-692,429, a novel non-peptidyl GH secretagogue. Endocrinology 1993;6:2729-2731.
41.Pong SS, Chaung L-YP, Dean D, Nargund RP, Patchett AA, Smith RG. Identification of a new G-protein-linked receptor for growth hormone secretagogues. Mol Endocrinol 1996;10:57-61.
42.Howard AD, Feighner SD, Cully DF, et al. Cloning of a G-protein coupled receptor for the growth hormone releasing peptide (GHRP-6) and non-peptidyl growth hormone secretagogues. Science 1996;273:974-977.
43.Sethumadhavan K, Veeraragavan K, Bowers CY. Demonstration and characterization of the specific binding of growth hormone-releasing peptide to rat anterior pituitary and hypothalamic membranes. Biochem Biophys Res Commun 1991;178:31-37.
44.Codd EE, Shu AYL, Walker RF. Binding of a growth hormone releasing hexapeptide to specific hypothalamic and pituitary binding sites. Neuropharmacology 1989;28:1139-1144.
45.Veeraragavan K, Sethumadhavan K, Bowers CY. Growth hormone-releasing peptide (GHRP) binding to porcine anterior pituitary and hypothalamic membranes. Life Sci 1992;50:1149-1155.
46.McKee KK, Palyha OC, Feighner SD, et al. Molecular analysis of rat pituitary and hypothalamic growth hormone secretagogues receptors. Mol Endocrinol 1997;11:415-423.
47.Bodart V, McNicoll N, Carriere P, et al. Identification and characterization of a new GHRP receptor in the heart. Program of the 80th Meeting of The Endocrine Society, New Orleans LO,1998:302.
48.Ong H, McNicoll N, Escher E, et al. Identification of a pituitary growth hormone-releasing peptide (GHRP) receptor subtype by photoaffinity labelling. Endocrinology 1998;139:432-435.

49. Berti F, Rossoni G, Magni G, et al. Nonsteroidal antiinflammatory drugs aggravate acute myocardial ischemia in the perfused rabbit heart: a role for prostacyclin. J Cardiovasc Pharmacol 1988;12:438-444.
50. De Gennaro Colonna V, Rossoni G, Bonacci D, et al. Worsening of ischemic damage in hearts from rats with selective growth hormone deficiency. Eur J Pharmacol 1996;314:333-338.
51. De Gennaro Colonna V, Rossoni G, Bernareggi M, Muller EE, Berti F. Cardiac ischemia and impairment of vascular endothelium function in hearts from GH-deficient rats: protection by hexarelin. Eur J Pharmacol 1997;334:201-207.
52. Lakatta EG, Yin FC. Myocardial aging: functional alterations and related cellular mechanisms. Am J Physiol 1982;242:H927-941.
53. Ji LL, Dillon D, Wu E. Myocardial aging: antioxidant enzyme systems and related biochemical properties. Am J Physiol 1991;261:R386-392.
54. Mudumbi RV, Olson RD, Hubler BE, Montamat SC, Vestal RE. Age-related effects in rabbit heart of N6-R-phenylisopropyladenosine, an adenosine A1 receptor agonist. Gerontology 1995;50A:B351-357.
55. Sonntag WE, Steger RW, Forman LJ, Meites J. Decreased pulsatile release of growth hormone in old male rats. Endocrinology 1980;107:1875-1879.
56. Rudman D, Kutner MH, Rogers CM, Lubin MF, Fleming GA, Bain RP. Impaired growth hormone secretion in the adult population. J Clin Invest 1981;67:1361-1369.
57. Torsello A, Luoni M, Grilli R, et al. Hexarelin stimulation of growth hormone release and mRNA levels in an infant and adult rat model of impaired GHRH function. Neuroendocrinology 1997;65:91-97.
58. Deghenghi R. Structural requirements of growth hormone secretagogues. In: Bercu BB and Walker RF (eds) Growth hormone secretagogues in clinical practice. Marcel Dekker Inc, New York, 1998:27-35.
59. Luoni M, Grilli R, Guidi M, et al. Effects of acute and long-term administration of growth hormone-releasing peptides on GH secretion and feeding behaviour in young-adult and aged rats. Program of the 79th Meeting of The Endocrine Society, Minneapolis MI, 1997:152.
60. Graziani A, Fubini A, Filigheddu N, et al. Hexarelin, a peptidyl GH secretagogue, is a novel survival factor for cardiomyocytes. J Endocrinol Invest 1999;22(suppl. to n. 4):16.
61. Valetto MR, Podio V, Broglio F, et al. Hexarelin, a peptidyl GH-secretagogue, increases the left ventricular ejection fraction but not GH secretion in adult patients with severe GH deficiency. Comparison with the effect in normal subjects. J Endocrinol Invest 1999;22(suppl. to n. 4):57.
62. Meissner A, Morgan JP. Contractile dysfunction and abnormal Ca2+ modulation during postischemic reperfusion in rat heart. Am J Physiol 1995;268:H100-111.
63. Tsukube T, McCully JD, Federman M, Krukenkamp IB, Levitsky S. Developmental differences in cytosolic calcium accumulation associated with surgically induced global ischemia: optimization of cardioplegic protection and mechanism of action. J Thorac Cardiovasc Surg 1996;112:175-184.
64. Cavero I, Boudot JP, Feuvray D. Diltiazem protects the isolated rabbit heart from the mechanical and ultrastructural damage produced by transient hypoxia, low-flow ischemia and exposure to Ca^{2+}-free medium. J Pharmacol Exp Ther 1983;226:258-268.
65. Nayler WG, Elz JS. Reperfusion injury: laboratory artifacts or clinical dilemma? Circulation 1986;74:215-221.

INDEX

Page numbers followed by "t" indicate tables

R

S

T

V

W